COLLECTIBLE NOVELTY PHONES

If Mr. Bell Could See Me Now...

James David Davis

Schiffer Publishing Ltd
4880 Lower Valley Rd. Atglen, PA 19310 USA

Library of Congress Catalog Card Number: 97-80683

DESIGNED BY BONNIE M. HENSLEY

ISBN: 0-7643-0472-0
Printed in China
1 2 3 4

Published by Schiffer Publishing Ltd.
4880 Lower Valley Road
Atglen, PA 19310
Phone: (610) 593-1777; Fax: (610) 593-2002
E-mail: schifferbk@aol.com
Please write for a free catalog.
This book may be purchased from the publisher.
Please include $3.95 for shipping.
Try your bookstore first.

We are interested in hearing from authors
with book ideas on related subjects.

DEDICATION

This book is dedicated to my wife, Elda, art director and proof-reader.

ACKNOWLEDGEMENTS

A million thanks go out to those wonderful people and friends who helped so much in making this a complete book. The project was made more interesting by their input.

The following furnished many phones and comments to help me along: Betty Chapman, Jim and Wynona Goodin, Timothy J. Eisenhauer, Jonathan Frank, and Ken King. I want to thank them again for all their support. I appreciate it very deeply.

Bill Berryman is what you call a photographer's photographer. He put in many hours behind the camera and always said, "If I didn't think this was a fun project, I wouldn't do it!" Bill's dedication and friendship made the whole project really special to my wife and me.

Another special thank you goes out to my daughter, Jill, for putting all the text into WordPerfect. This was a laborious task with many late night corrections.

Thanks to Peter Schiffer, publisher, for encouraging us to finish this project.

CONTENTS

INTRODUCTION

Many dedicated telephone collectors shun the thought of collecting or even keeping the funny, odd-shaped, silly looking novelty or character phones. As older phones become scarce and more costly, increasing numbers of collectors are drawn to the glitter, the bright colors, the weird figures, or the company logos on the unusual, novelty phones, whether old or new.

The earliest novelty phones had limited feature options such as touch tone or pulse dialing keypads and a ringer hi/low volume control switch. The later phones have those features plus many added ones. Today's features include a mute key to prevent private conversation from being transmitted to the person called; a last number redial key; a lighted keypad dial; and an out-pulse, push-button keypad system which works as a touch-tone or rotary dial. Some phones make a sound other than a ring, or lights flash to signal an incoming call. After the breakup of the Bell System, telephone manufacturers began to include FCC rules, requirements, and precautions with their phone instructions. As the reader will notice, many of the telephones were manufactured outside of the United States.

In the past, telephone collector books dealt with antiques, or the design changes of the telephone through the years, with little mention of novelty phones. This book is intended as a reference guide for old and new collectors, enabling them to become more knowledgeable on the subject.

This book contains five chapters: Character Telephones, Advertising Telephones, Figural Telephones, Miniatures, and Toy Telephones. A color print of each phone is included along with a description of each phone's characteristics and specifications.

The prices for the novelty phones in this publication are based on current periodicals and prices found in flea markets, and antique shops. Prices will vary across the country. Some of these phones are becoming very scarce and quite expensive.

So happy hunting and enjoy the book!

Chapter I

WHAT A CHARACTER

Character Telephones

The popularity of the Bell System Mickey Mouse and Snoopy telephones was largely responsible for increased interest in other character and novelty type phones being produced today. Along with Kermit and Winnie the Pooh, these phones were the forerunners of today's novelty telephones. They were produced to commemorate the United States' bicentennial in 1976.

Character collecting is hot. Superman, Popeye, Pillsbury Doughboy, Elvis, and the ever-present Mickey Mouse are some of the most popular characters in the present day market. Character Telephones are found in a variety of mediums, such as comic books, comic strips, cartoons, movies, music, and television. Some of these phones could be considered advertising, but seem to fit better into a character category.

The Alien Disc Telephone was manufactured in China for Telemania Products and distributed by Kash N' Gold Ltd., Ronkonkoma, New York. Model: Alien Phone. 1997. 3.25" deep x 11.75" diameter. Touch-tone or pulse-dialing switch on side of handset. Ringer on/off switch on back. Bell tone hi/low switch on back. Hearing aid compatible. Last-number redial key. Mute key. Flash key for new on/off hook. Keypad, keys, transmitter, and receiver on handset. On/off hook switch in handset cradle. Desk set or wall mount. $30-40.

The Alf Telephone was made in Hong Kong. Model: 618S. Circa 1980s. 10.5" H x 9.5" W x 8" deep. Touch-tone or pulse-dialing switch. Ringer on/off switch. Last-number redial key. Keypad, key, switches, on/off hook switch, transmitter, and receiver on handset. Desk set. $80-110.

The Alvin and the Chipmunks Telephone was manufactured in Hong Kong for Bagdasarian Productions licensed by Mode Productions, New York. 1984. 15" H x 6.5" W 6" deep. Touch-tone or pulse-dialing switch. Mute key. Last-number redial key. Ringer on/off switch. Keypad, switches, keys, transmitter, receiver, and on/off hook switch on handset. Desk set. $60-90.

The Barbie® 'Solo in the Spotlight' Telephone was manufactured in China for Segan Products and designed by M. H. Segan, L.P. Distributed by Kash N' Gold. Model: Barbie® Telephone. 1995. 12.5" H x 9.25" W x 8.25" deep. Touch-tone or pulse-dialing switch. Mute key. Last-number redial key. Requires three C-size batteries. Battery compartment under base. Light flashes and sings original 1959 "Busy Bee" soundtrack and stage light flashes when phone rings. Doll is made of cold-cast porcelain. Ringer on/off switch. Ringer or Barbie® singing switch. Transmitter and receiver on handset. Switches and speaker underneath base. Keypad, keys, and on/off hook switch on top of base. Desk set. $65-95.

The Bart Simpson Telephone was manufactured in Taiwan for Columbia TEL-COM Columbia Telecommunications Group Incorporated. Under license from Twentieth Century Fox Film Corp. 1990. 8" H x 3.75" W x 4" deep. Touch-tone or pulse-dialing switch. Mute key. Last-number redial key. On/off hook switch underneath base. Keypad, keys, switch, transmitter, and receiver on back. Desk set. $30-40.

Opposite page:
Top: The Barbie® Telephone was manufactured in Hong Kong for Fun Communications under license from Mattel Incorporated. 1983. 8" H x 8" W 8" deep. Touch-tone dialing. Mute key. Last-number redial key. Keypad, keys, on/off hook switch, transmitter, and receiver on handset. Desk set. $150-180.

The Blossom Television Show Telephone was manufactured in China for C. Williams Corp. 1993. 2" H x 2" W x 8" deep. Touch-tone dialing. Ringer on/off switch. On/off switch located on side of handset. In-use LED indicator light on back of handset. Keypad, switch, on/off hook switch, transmitter, and receiver on handset. Desk set or wall mount with hanger on top of handset. $50-70.

The Batman Batmobile Car I Telephone was manufactured in China for DC Comics Incorporated. 1989. 2.5" H x 4" W x 10" deep. Touch-tone or pulse-dialing switch. Mute key. Last-number redial key. Ringer on/off switch. Keypad, keys, switches, on/off hook switch, transmitter, and receiver underneath car. Desk set. $60-80.

The Batman Batmobile Car II Telephone was manufactured in China for Micro Games of America under license from DC Comics Incorporated. Model: BAT-8040. 1995. 3.75" H x 3.5" W x 11.75" long. Touch-tone or pulse-dialing switch. Last-number redial key. Flash on/off key for new dial tone. Lights up in blue when phone rings. Keypad, keys, switch, on/off hook switch, transmitter, and receiver underneath car. Desk set. (*Courtesy of Betty C.*) $40-60.

The Batman Tennis Shoe Telephone was manufactured in China for DC Comics Incorporated. 1989. 6.5" H x 4" W x 12" deep. Touch-tone dialing. Mute key. Last-number redial key. Ringer on/off switch. Keypad, keys, switch, on/off hook switch, transmitter, and receiver on bottom of shoe. Desk set. $70-95.

The Betty Boop Soft Telephone was manufactured in China by Reyco Incorporated. Designed by Nicolai Canetti. 1992. 1.75" H x 3" W x 8.5" deep. Touch-tone or pulse-dialing switch. Last-number redial switch. Ringer on/off switch. In-use LED indicator light on handset. On/off power key (on/off hook switch). Keypad, switches, key, transmitter, and receiver on handset. Desk set. $55-75.

The Big Bird Telephone was manufactured in China for Muppets Incorporated. Model: 1725. Circa 1990s. 6.75" H x 4.25" W x 4.5" deep. Touch-tone dialing. Last-number redial switch. Mute key. On/off hook switch underneath bird. Keypad, switch, key, transmitter, and receiver on back of bird. Desk set. $55-75.

The Bozo the Clown Telephone was manufactured in Thailand for Kash N' Gold Ltd., New York, distributors for Telemania Products. Model: Bozo Phone. 1988. 11.5" H x 6" diameter. Touch-tone or pulse-dialing switch. Ringer on/off or Bozo laughing switch underneath base by battery compartment. Last-number redial switch. Requires four AA batteries. Remove clown from base to answer or dial. On/off hook switch on top of base. Keypad, switches, transmitter, and receiver located on back of clown. Desk set. $60-80.

The Beetle Bailey Telephone was manufactured in Hong Kong for Comvu Corp. 1982. 9.75" H x 3.25" W x 2.25" deep. Touch-tone dialing. Keypad and on/off hook switch underneath base. Transmitter and ringer on/off switch on top of base. Receiver on front of barrel. Desk set. $85-125.

The Bugs Bunny Telephone was manufactured in Hong Kong for Comvu Corp. 1982. 12" H x 4" W x 4.5" deep. Touch-tone dialing. Keypad and on/off hook switch underneath base. Transmitter and ringer on/off switch on top of base. Receiver on front of barrel. Desk set. $100-130.

The Bugs Bunny Relaxing Telephone was manufactured in China for Toshiba American Consumer Products Incorporated, New Jersey. Toshiba Looney Tunes Tronics product line. Model: LT800BUG. 1996. 7.25" H x 4" W x 12.25" length. Touch-tone or pulse-dialing switch. Last-number redial key. In-use LED indicator. Ringer on/off switch. Flash key for new on/off hook switch. On/off hook switch under Bug's seat. Keypad, switches, keys, and LED on top of base. Transmitter is Bug's foot. Receiver is Bug's mouth. Desk set. $60-80.

The Buzz Lightyear Telephone was manufactured in China for Brooktel under license by Lenbrook Industries Ltd. Model: Buzz Lightyear Buzz9609. 1996. 15" H x 12" W x 9" deep. Touch-tone or pulse-dialing switch. Last-number redial key. Ringer on/off switch. Blast off light rocket flame lights up when phone rings. Hearing aid compatible. Flash on/off key. Keypad, keys, switches, transmitter, and receiver on handset. On/off hook switch on handset cradle. Desk set. $65-85.

The Cabbage Patch Kids Telephone was manufactured in Hong Kong for Coleco, Hartford, Connecticut. Model: 7911. 1984. 13.25" H x 7.75" W x 8.75" deep. Touch-tone dialing. Mute key. Last-number redial key. Ringer on/off switch on back of base. In-use LED indicator light on top of base. On/off hook seems to be magnetically controlled between handset and arm. Keypad and keys on top of base. Transmitter and receiver on handset. Desk set. $130-170.

The Elvis Presley Telephone was manufactured in China for Telemania Products and distributed by Kash N' Gold Ltd., Ronkonkoma, New York. The phone will sing "Jail House Rock" and dance in time with music when the phone rings. Model: Elvis Telephone. 1996. 14.5" H x 8.25" W x 10" deep. Touch-tone or pulse-dialing switch on handset. Ringer or Elvis dancing and singing switch underneath base. Requires three D-size batteries in compartment underneath base. Hearing aid compatible. Star and pound keys for various purposes on long distance or other tone/dial systems applications. Last-number redial key on handset. Keypad, transmitter, receiver, and on/off hook switch on handset. Desk set. $60-85.

The Cookie Monster of Sesame Street Telephone was manufactured in China for Children's Television Workshop Sesame Street by Concept 2000, a division of Justin Products, New York. 1986. 8" H x 3.5" W x 3" deep. Touch-tone dialing. Mute key. Last-number redial key. On/off hook switch underneath base. Keypad, keys, transmitter, and receiver on back. Desk set. $120-150.

The Eric Carmen 'Hot Lips' Telephone was made by Telequest. This was a not-for-sale promotional for a record of Eric Carmen's "I Wanna Hear It from Your Lips" produced by Green Records. 1984. 4.5" H x 8.5" W x 2.5" deep. Touch-tone or pulse-dialing switch. Last-number redial key. Keypad, key, switch, on/off hook switch, transmitter, and receiver located on handset top part of lips. Desk set. (*Courtesy of Jim and Wynona Goodin.*) $65-95.

The Fortune Mouse Telephone is licensed and manufactured by American Canyon Ltd. Model: ACL-1996. 1996. 11.5" H x 9.5" W x 5.25" deep. Touch-tone or pulse-dialing switch. Last-number redial key. In-use LED indicator. Ringer on/off switch behind the telephone address label. Can be used as a savings box. Slot on back of mouse's head. Keypad, switch, key, LED, transmitter, receiver, and on/off hook switch on handset. Desk set. $75-100.

The Garfield Climbing Wall Telephone was manufactured in Taiwan by May Yung Enterprises Co. for Tyco Industries Incorporated, Moorestown, New Jersey. Model: 1209. Circa 1980s. 12" H x 9.5" W x 6" deep. Touch-tone or pulse-dialing switch. Mute key. Last-number redial key. Keypad, keys, switch, transmitter, receiver, and on/off hook switch on handset. Desk set. $65-100.

The Garfield Lighting and Music Sound Telephone was licensed and manufactured in Europe by American Canyon Ltd. Model: ACL-1238. Circa 1990s. 9.5" H x 7.25" W x 6.5" deep. Touch-tone or pulse-dialing switch. Last-number redial key. Ringer on/off or music switch. Light on/off switch. LED in-use indicator. Requires three AAA batteries. Speaker and battery compartment underneath base. When phone rings light flashes and music sounds. Keypad, keys, switches, transmitter, receiver, and on/off hook switch on handset. Desk set. (*Courtesy of Timothy J. Eisenhauer and Jonathan Frank.*) $40-65.

The Garfield Lying Down Small Telephone is licensed and manufactured by American Canyon Ltd. Model: ACL-1208. 1995. 3.5" H x 5.5" W x 9" length. Touch-tone or pulse-dialing switch. Last-number redial key. Hearing aid compatible. Keypad, switch, transmitter, receiver, and on/off hook switch underneath Garfield. Desk set. $40-60.

The Garfield Lying Down Large Telephone was manufactured for Tyco Industries Incorporated, Moorestown, New Jersey, by S. Megga Telecommunications Ltd. 1978. 5" H x 7.25" W x 10.5" length. Touch-tone dialing. Ringer on/off switch. Keypad, switch, on/off hook switch, transmitter, and receiver on handset on back of Garfield. Desk set. $35-60.

The Garfield Talking Telephone was manufactured in China by Core enterprises for Tyco Industries Incorporated, Moorestown, New Jersey. Model: 1208. 1981. 9.5" H x 8" W x 7.25" deep. Touch-tone or pulse-dialing switch. Last-number redial key. Requires three AA batteries. Speaker and battery compartment underneath base. When phone rings, Garfield speaks "Hold my calls," "I'm not in," "Yo phone," "Yoo hoo," "Hello," "It's the President," "It's the Governor," or "It's the phone." Keypad, key, switch, transmitter, receiver, and on/off hook switch on handset. Desk set. $75-100.

The Marvin the Martian Telephone was made in China by Toshiba for American Consumer Products Incorporated, New Jersey. Toshiba Looney Tunes Tronics product line. Model: LT800MAR. 1997. 8.5" H x 7.75" W x 7.5" deep. Touch-tone or pulse-dialing switch. Last-number redial key. Flash or mute key. Tone key for new on/off hook. In-use LED indicator. Ringer on/off underneath base. Keypad, switch, keys, transmitter, and receiver on handset. On/off hook switch in handset cradle. Desk set. (*Courtesy of Betty C.*) $40-60.

The Ghostbusters Hard Plastic Telephone was made in Hong Kong by Galax for Remco Toys Incorporated. 1987. 7.5" diameter. Touch-tone dialing. Mute key. Last-number redial key. Ringer on/off switch. Keypad, keys, switch, transmitter, receiver, and on/off hook switch on handset. Desk set or wall mount. (*Courtesy of Timothy J. Eisenhauer and Jonathan Frank.*) $125-140.

Opposite page:
The Darth Vader Speaker Telephone was manufactured in Hong Kong by American Telecommunications Corp. under permission of Lucas Film Ltd. Model: DEIF 8040. 1983. 14" H x 6.5" W x 8" deep. Touch-tone dialing. Mute key. Last-number redial key. On/off hook switch on base. Volume control, speaker, keypad, and keys on top of base. Desk set. $120-150.

The Ghostbusters Thin Plastic Telephone with artwork and design by RCA Columbia Pictures Home Video. Model: ST-738. 1985. 10.75" H x 12" W x 3" deep. Touch-tone dialing. Mute key. Last-number redial key. Keypad, keys, transmitter, receiver, and on/off hook switch on handset. Desk set or wall mount. $125-140.

1 2 3 4 5
MIC MUTE 6 7 8 9 0

The Gumby Telephone. Circa 1980s. 5" H x 4" W x 9.75" deep. Touch-tone dialing. Mute key. Last-number redial key. Keypad, keys, transmitter, and receiver on handset. On/off hook switch underneath Gumby. Desk set. $125-150.

The He-Man Masters of the Universe Telephone was made in Hong Kong for Trans-World Sales Incorporated. Model: 8530. 1983. 8" H x 8" length. Touch-tone dialing. Mute key. Last-number redial key. Keypad, keys, transmitter, receiver, and on/off hook switch on handset. Desk set. (*Courtesy of Timothy J. Eisenhauer and Jonathan Frank.*) $100-130

The Hagar the Horrible Telephone was manufactured in Hong Kong for Comvu Corp. 1982. 10" H x 9" length. Touch-tone dialing. Keypad on base. On/off hook switch in handset cradle. Transmitter and receiver on handset. Desk set. (*Courtesy of Timothy J. Eisenhauer and Jonathan Frank.*) $350-500.

The Kermit the Frog Telephone was manufactured in Singapore by Henson Associates Incorporated for American Telecommunications Corp. Distributed by the Bell System. These were available with a lamp attachment. Model: TEIF 8030. 1983. 11.5" H x 9" W x 8.5" deep. Touch-tone dialing. Ringer hi/low switch under base. Keypad on top of base. On/off hook switch is Kermit's leg. Transmitter and receiver on handset. Desk set. $140-175.

The Mighty Mouse Telephone was made in Taiwan for John Stanley Industries, New York. 1983. 9" H x 6.75" W x 1.75" deep. Touch-tone dialing. Mute key. Last-number redial key. Keypad, keys, switch, on/off hook switch, transmitter, and receiver on handset. Desk set or wall mount. $65-95.

Above: The Kermit Candlestick Telephone was manufactured in China for Telemania Products and distributed by Kash N' Gold Ltd., Ronkonkoma, New York. A Segan product created and designed by M.H. Segan. L.P. Kermit, trademark of Jim Henson Productions, Incorporated. Model: Kermit Telephone. 1996. 12.5" H x 5.5" diameter. Touch-tone or pulse-dialing switch underneath base. Ringer on/off switch underneath base. Hearing aid compatible. Star and pound keys for various purposes on long distance or other tone/dial systems applications. Last-number redial key and keypad on top of base. Transmitter is Kermit's mouth. Handset is receiver. On/off hook switch is Kermit's arms. Desk set. $45-75.

Right: The Nickelodeon Talk Blaster Telephone was manufactured in China by Long Hall Technologies for Viacom International Incorporated. Model: N2500. 1996. 7" H x 8" W x 7" deep. Touch-tone or pulse-dialing switch on back of base. Last-number redial key. Flash key for new on/off hook. Requires three C-size batteries. Battery compartment under base. Ringer or Nickelodeon four fun sounds: Nick noise, honk, moo, and laugh switch on back of base. Green slime power rod lights up when phone rings. Ringer hi/low volume switch. Hearing aid compatible. Keypad, keys, and on/off hook switch, transmitter, and receiver on handset. Desk set. $60-85.

The Mickey Mouse Telephone was manufactured in China for the A.T. & T. Co. and distributed by many retail stores. Circa 1990s. 14.5" H x 10" W x 9" deep. Touch-tone or pulse-dialing switch. Mute key. Last-number redial key. Ringer off/hi/low switch in phone holder. Keypad, switch, keys, transmitter, receiver, and on/off hook switch on handset. Lighted dial. Desk set. $50-80.

The Mickey Mouse Telephone was manufactured by American Telecommunications Corp. under permission of the Disney Co. and was distributed by the Bell System. These were available with a lamp attachment with choice of touch-tone pad or rotary dial. 1976. 15" H x 9.5" W x 8" deep. Touch-tone dialing. Ringer hi/low volume switch underneath base. Keypad on top of base. Transmitter and receiver on handset. Mickey's arm is on/off hook switch. Desk set. $60-100.

The Mickey Mouse, Bell System Telephone with lamp attachment. $60-100.

The Mickey Mouse with Blue Backpack Telephone was manufactured in Hong Kong by Tyco Industries Incorporated. Model: 1245. 1987. 14" H x 6.25" W x 7" deep. Touch-tone or pulse-dialing switch. Mute key. Last-number redial key. Keypad, switch, keys, transmitter, receiver, and on/off hook switch on backpack handset. Desk set. $75-100.

Opposite page:
The Mickey Mouse Round Base Telephone was manufactured in China by Synanon for the Disney Co. 1988. 10.5" H x 4.75" diameter. Touch-tone or pulse-dialing switch. Mute key. Last-number redial key. Keypad, switch, keys and on/hook switch underneath base. Transmitter behind head. Receiver in back on top of base. Desk set. $140-175.

The Mickey Mouse Message Pad Telephone was manufactured in Korea by Unisonic Products Corp. for Walt Disney Productions Incorporated. Model: TP 6050. 1988. 9.75" H x 8.75" W x 5.75" deep. Touch-tone dialing. Mute key. Last-number redial key. LED power indicator light. Ringer on/off switch. Erasable message pad panel. Slot for holding felt markers on base. Keypad, switch, keys, LED, transmitter, receiver, and on/off hook switch on handset. Desk set. $60-90.

The Mickey Mouse Soft Telephone was manufactured in China by Reyco Incorporated. Designed by Nicolai Canetti. 1992. 8.25" H x 2.85" W x 1.75" deep. Touch-tone or pulse-dialing switch. Last-number redial switch. Ringer on/off switch. In-use LED indicator light on handset. On/off power key (on/off hook switch). Keypad, switches, key, transmitter, and receiver on handset. Desk set. $45-70.

The Orca the Whale Telephone was made in Taiwan for Telemania Products and distributed by Kash N' Gold Ltd., Ronkonkoma, New York. Model: HAC-KW-888. Circa 1990s. 9.75" H x 5.5" W x 9" deep. Touch-tone or pulse-dialing switch. Orca makes sounds of splashing, cries, and wails. Requires 9V battery compartment underneath. Last-number redial key. Ringer or whale sound switch. Keypad, on/off hook switch, switches, key, transmitter, and receiver are located on whale's belly. Desk set. (*Courtesy of Betty C.*) $50-65.

The Mickey Mouse Standing by Tree Stump Telephone is licensed and manufactured by American Canyon Ltd. Model: ACL-M3. 1996. 8.5" H x 4.25" W x 4.25" deep. Touch-tone or pulse-dialing switch. Last-number redial key. In-use LED indicator. Ringer on/off switch. On/off button is on/off hook switch. Keypad, switches, keys, LED, transmitter, receiver, and on/off button on back of tree stump. Desk set. $75-95.

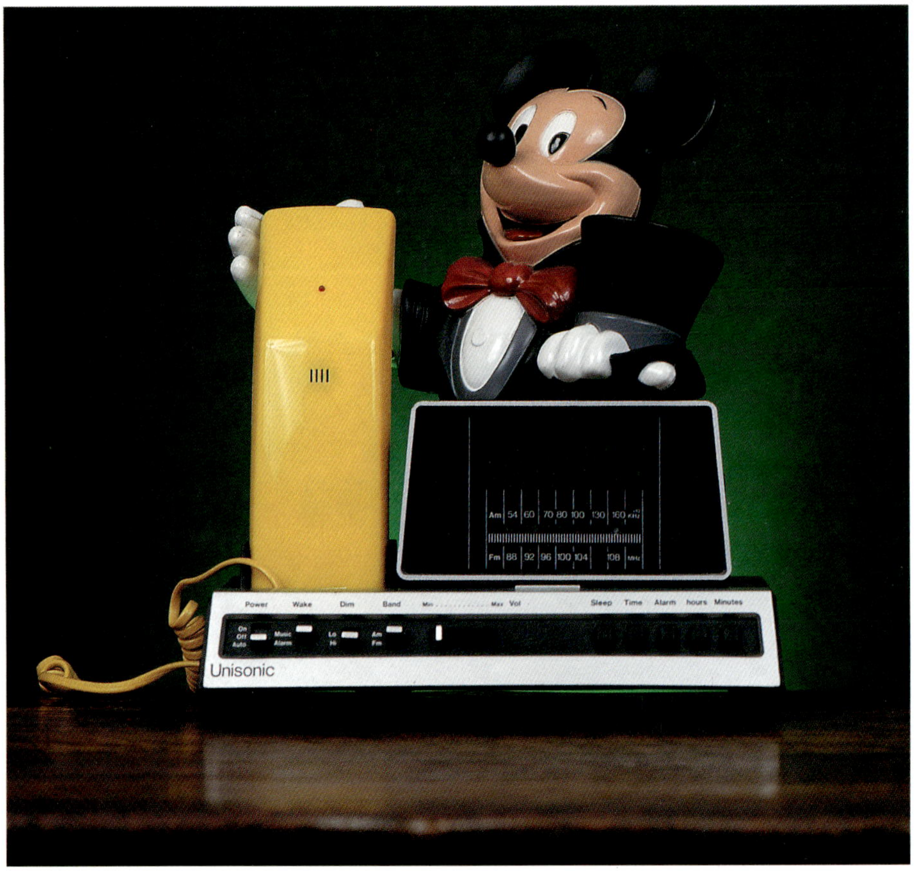

The Mickey Mouse Clock Radio Telephone was manufactured in Hong Kong for Unisonic Products Corp. for Walt Disney Productions Incorporated. Model: TP 6052. 1983. 12" H x 10" W x 6.75" deep. Touch-tone dialing. Mute key. Last-number redial key. Ringer on/off switch. In-use LED indicator light. Keypad, switch, keys, transmitter, receiver, LED, and on/off hook switch on handset. Desk set. $70-100.

Above: The New Kids on the Block Telephone was made in China for Big Step Productions Incorporated and distributed by Hasbro Incorporated, Pawtucket, Rhode Island. 1990. 8" H x 2 .5" W x 1.75" deep. Touch-tone or pulse-dialing switch. Mute key. Last-number redial key. Keypad, keys, switch, on/off hook switch, transmitter, and receiver on handset. Desk set or wall mount. $130-160.

Right: The Pacman Telephone was manufactured in Hong Kong for American Telecommunications Corp. under license of 80 Bally Midway Manufacturing Co. Model: DEIF-8410. Circa 1980s. 2" H x 4.5" diameter. Touch-tone dialing. Mute key. Last-number redial key. Keypad, keys, transmitter, receiver, and on/off hook switch inside lid. Desk set. $45-70.

The Opus Penguin Telephone was manufactured in Hong Kong for Tyco Industries Incorporated under license of the Washington Post Writers Group. 1987. 13" H x 7.5" W x 8" deep. Touch-tone or pulse-dialing switch. Mute key. Last-number redial key. Head swivels when handset is picked up. Keypad, switch, keys, transmitter, and receiver, on handset located on back. On/off hook switch in handset compartment. Desk set. $120-150.

The Orphan Annie and Sandy Sitting Telephone was manufactured by GTE Automatic Electric under license from Tribune Co. Syndicate Incorporated. 1981. 7" H x 2.75" W x 9" deep. Pulse dialing. Rotary dial, transmitter, and receiver on handset. On/off hook switch in base cradle. Desk set or wall mount. $150-180.

The Orphan Annie and Sandy Standing Telephone was manufactured in Hong Kong for the Comvu Corp. 1982. 10.5" H x 3" W x 4.5" deep. Touch-tone dialing. Keypad and on/off hook switch underneath base. Receiver located in back on backpack. Transmitter located in back on Sandy. Desk set. $125-150.

The Power Rangers Mighty Morphin Telephone was manufactured in China for REC Sound Incorporated under license of Saban Entertainment Incorporated. Model: PRT-300. 1994. 8.75" H x 9.25" W x 9.25" deep. Touch-tone or pulse-dialing switch. Mute key. Last-number redial key. In-use LED indicator light. Flash key for call waiting. Ringer on/off switch. Talk on/off hook switch. Keypad, switches, keys, LED, transmitter, and receiver on handset. Desk set. $60-85.

The Robo Force Telephone was manufactured in Hong Kong. Model: 7760. 1984. 12" H x 7.5" W x 6" deep. Touch-tone dialing. Mute key. Last-number redial key. Ringer on/off switch. In-use LED indicator light. Transmitter and receiver on backside of handset. Keypad, switch, keys, on/off hook switch, and LED on handset. Desk set. $100-130.

The Robot SR006 Telephone was manufactured in Taiwan by Fabrique. Circa 1980s. 10" H x 5" W x 6" deep. Touch-tone or pulse-dialing switch. Pause key same as mute key. Last-number redial key. Reset key for new dial tone. Keypad, switch, keys, on/off hook switch, transmitter, and receiver on handset. Desk set. $60-90.

The Popeye Sitting and Sweet Pea Telephone was made by Imperial. Circa 1980s. 17" H x 8" W. Touch-tone dialing. Keypad on top of base. On/off hook switch arm. Transmitter and receiver on handset. Desk set. (*Courtesy of Timothy J. Eisenhauer and Jonathan Frank.*) $450-600.

The Popeye Standing Telephone was manufactured in Hong Kong for Comvu Corp. 1982. 9.5" H x 4.75" W x 5" deep. Touch-tone dialing. Keypad and on/off hook switch underneath base. Receiver is located on front of spinach can. Transmitter is located under duffel bag on base. Desk set. $95-150.

The Rolling Stones Tongue Telephone was made in Hong Kong by Musidor B.V. for Tristar International Ltd. 1983. 3" H x 5.5" W x 8.5" length. Touch-tone dialing. Mute key. Last-number redial key. Ringer on/off switch. Keypad, keys, switch, on/off hook switch, transmitter, and receiver on tongue handset. Desk set. $250-350.

The Snoopy and Woodstock Telephone was manufactured by American Telecommunications Corp. under permission of the Disney Co. and distributed by the Bell System. These were available with a lamp attachment with choice of touch-tone pad or rotary dial. 1976. 13" H x 9.5" W x 8" deep. Touch-tone dialing. Hi/low volume switch underneath base. Keypad on top of base. Transmitter and receiver on handset. Snoopy's arm is on/off hook switch. Desk set. $70-110.

The Snoopy Lying Down or Standing Telephone was made for United Syndicate Incorporated. Circa 1980s. 7.5" H x 3.25" W x 4.75" deep. Touch-tone dialing. Mute key. Last-number redial key. Keypad, keys, transmitter, receiver, and on/off hook switch on back of Snoopy. Desk set or wall mount with bracket. $40-75.

The Joe Cool Snoopy and His Pal Woodstock Telephone was manufactured in China by Seika Corp. Model: SLK-291. 1993. 11" H x 8.5" W x 8.5" deep. Touch-tone or pulse-dialing switch. Last-number redial key. Hold key. LED indicator light shows when call is on hold. Ringer on/off switch. Keypad, switches, keys, transmitter, receiver, LED, and on/off hook switch on handset. Desk set. $80-110.

The Snoopy with Bank Telephone was made in Hong Kong for American Canyon Ltd. Model: ACL-288. 1995. 9" H. Touch-tone dialing. Ringer on/off switch. Last-number redial key. In-use LED indicator. Lighted handset. Keypad, key, switch, transmitter, receiver, and on/off hook switch on handset. Desk set. (*Courtesy of Timothy J. Eisenhauer and Jonathan Frank.*) $60-80.

The Spider-Man Telephone was manufactured in Hong Kong for REC Sound Incorporated under permission of Marvel Entertainment Group Incorporated. This was the second Spider-Man phone manufactured. Model: SST-200. 1994. 9.5" H x 9.5" W x 9" deep. Touch-tone or pulse-dialing switch. Mute key. Last-number redial key. In-use LED indicator light. Flash key for call waiting. Ringer on/off switch. Talk on/off hook switch. Keypad, switches, keys, LED, transmitter, and receiver on handset. Desk set. $60-85.

The Amazing Spider-Man Telephone was manufactured in Hong Kong by MCG. This was the first Spider-man phone manufactured. 1984. 11" H x 9.5" W x 8.5" deep. Touch-tone dialing. Mute key. Last-number redial key. Ringer on/off switch. Keypad, switch, keys, on/off hook switch, transmitter, and receiver on handset. Desk set. $100-140.

The Star Trek the Next Generation Telephone was manufactured in China for Telemania Distributed by Kash N' Gold under license of Paramount Pictures. Model: Star Trek The Next Generation Telephone. 1996. 9.25" H x 12" W x 12" deep. Touch-tone or pulse-dialing switch on handset. Ringer on/off switch underneath base. Mode switch underneath base. Electronic ringer by Soundtrax or Star Trek "Cone Breach" sound. Lighted Enterprise Ship and Red Planet flash when phone rings. Hearing aid compatible. Keypad, transmitter, and on/off hook switch on handset. Last-number redial key on handset. Desk set. $50-70.

The Star Trek USS Enterprise Ship Telephone was manufactured in China for Telemania. Distributed by Kash N' Gold under license of Paramount Pictures. Model: Startrek. 1993. 7.5" H x 6.5" W x 12.5"Length. Touch-tone or pulse-dialing switch underneath base. Ringer or red alert sounds and "Warp Nacelles" flash when phone rings. Switch underneath base. Keypad on top of base. Transmitter and receiver located in handset on top half of ship. Two on/off hook switches, one H key on top of base and the other inside ship. Last-number redial key on top of base. Desk set. $60-80.

Left: The Super Mario Brothers Telephone was manufactured in China for Dialfone Ltd. 1992. 5" H x 2.5" W x 8" length. Touch-tone or pulse-dialing switch. Mute key. Last-number redial key. Ringer on/off switch. Mario swivels. Keypad, switches, keys, transmitter, receiver, and on/off hook switch underneath. Desk set. $65-100.

Above: The X-Men Wolverine Telephone was manufactured in China for REC Sound Incorporated under permission of Marvel Entertainment Group Incorporated. Model: KWT-100. 1994. 8.5" H x 8.75" W x 7.5" deep. Touch-tone or pulse-dialing switch. Mute key. Last-number redial key. In-use LED indicator light. Flash key for call waiting. Ringer on/off switch. Talk on/off hook switch. Keypad, switches, keys, LED, transmitter, and receiver on handset. Desk set. $60-85.

Right: The Tom and Jerry the Movie Telephone was manufactured in China by Ronsonic Trading Corp. for Turner Entertainment Co. and Telefilm Essen Gmbh. Model: HAC RON-115. 1993. 11" H x 2.75" W x 2.5" deep. Touch-tone or pulse-dialing switch. Mute key. Last-number redial key. Keypad, switch, keys, transmitter, receiver, and on/off hook switch on back. Desk set. $85-125.

The Superman Handset in Back on Base Telephone was manufactured in China for Allied Telecommunications. About 300 were produced before DC Comics Incorporated stopped Co. for trademark violations. Rare. 1978. 17.75" H x 9.75" W x 9.75" deep. Pulse dialing. Dial in front on top of base. On/off hook switch, transmitter, and receiver on handset. Desk set. (*Courtesy of Jim and Wynona Goodin.*) $900-1,200.

Opposite page:
Back view of Superman with handset in back on base.

The Superman Handset in Back on Hard Plastic Cape Telephone was manufactured in China for ATE Electric under license from DC Comics Incorporated. 1979. 17.75" H x 8.25" W x 9.5" deep. Pulse dialing. Dial in front on top of base. On/off hook switch in back on cape. Transmitter and receiver on handset in back on cape. Desk set. $900-1,200.

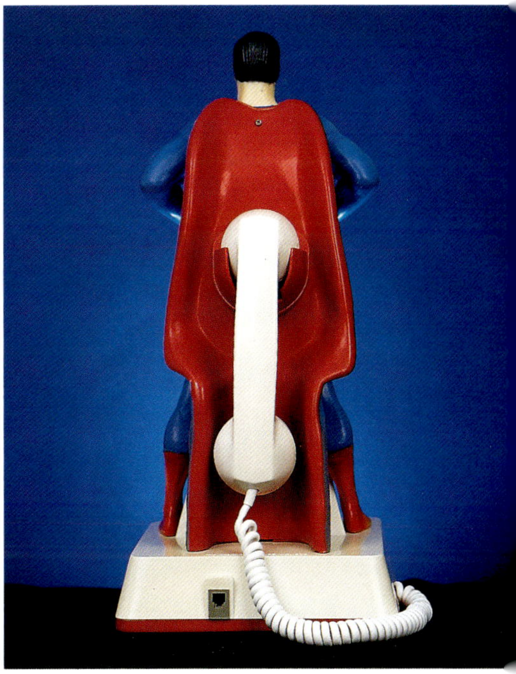

Back view of Superman with handset on back of cape.

The Winnie the Pooh Telephone was manufactured by American Telecommunications Corp. under permission of the Disney Co. and distributed by the Bell System. These were available with a lamp attachment. 1976. 14.25" H x 10.25" W x 8" deep. Touch-tone dialing. Hi/low volume switch underneath base. Keypad on top of base. Transmitter and receiver on handset. Winnie's arm is on/off hook switch. Desk set. $180-225.

The Ziggy and Friends Telephone was manufactured in China for Tyco Industries Incorporated, Moorestown, New Jersey. Model: 1268. 1990. 8" H x 12" W x 8" deep. Touch-tone or pulse-dialing switch. Last-number redial key. Keypad, switch, key, transmitter, receiver, and on/off hook switch on handset. Desk set. $85-125.

Chapter II

MAD FOR ADS

Advertising Telephones

In the early 1980s, with the Bell System break-up, a multitude of manufacturers began to produce inexpensive telephones. Many of these phones were used to promote company products or logos and university and professional sports programs. These phones were awarded to company personnel, given as complimentary gifts or promos to retail dealers, offered in magazine mail-ins, or sold through company gift shops.

Above: The AC Spark Plug Telephone was manufactured in Taiwan by Synanon. 1985. 10" H x 3" diameter. Touch-tone dialing. Transmitter and receiver located on front of plug. Keypad and on/off hook switch underneath base. Desk set. $80-110.

Below: The Avon Telephone was manufactured in Malaysia for Unisonic under license of Avon Corp. Model: AV-52. Circa 1990s. 3.75" H x 3" W x 8.75" length. Touch-tone or pulse-dialing switch. Last-number redial key. Mute key. Reset key. Flash key. Keypad, switch, keys, transmitter, and receiver on handset. On/off hook switch in handset cradle. Desk set or wall mount. $40-60.

Below: The Bill Elliot Race Rig Fone McDonald's Semi-Truck Telephone was made in China for Columbia TEL-COM Group Inc., New York, under license from McDonald's Corp. Model: Truck-94. 1997. 3.75" H x 2.25" W x 14" length. Touch-tone or pulse-dialing switch. Last-number redial key. Mute key. Keypad, keys, switch, on/off hook switch, transmitter, and receiver underneath truck. Truck has room to carry two racing cars, extra engines, and all the team's equipment. Authentically styled Kenworth tractor pulls the rig. Desk set. (*Courtesy of Jim and Wynona Goodin.*) $30-45.

Left: The Brut Champagne Telephone was made in Taiwan for Maxim Hotel Casino. Circa 1980s. 10.75" H x 2.75" diameter. Touch-tone dialing. Last-number redial key. Mute key. Keypad, keys, switch, and on/off hook switch, underneath bottle. Transmitter and receiver on front of bottle. Desk set. (*Courtesy of Betty C.*) $25-35.

Right: The Budweiser Beer Big Can Telephone was manufactured by Paul Nelson Industries, Englewood, Colorado. Circa 1980s. 12.75" H x 8.5" W x 6.5" diameter. Touch-tone dialing. Transmitter and receiver on handset. On/off hook switch on side of can. Handset Stromberg-Carlson. Keypad on top of can. Desk set. $140-180.

The Budweiser Beer One Airship Telephone was made in China for Ronsonic Trading Corp., Old Bridge, New Jersey, under license from Anheuser-Busch Incorporated. Model: RON-135. 1997. 5" H x 3.75" W x 10.5" length. Touch-tone or pulse-dialing switch. Last-number redial key. Flash key for on/off hook. Pause or mute key. Ringer on/off switch. Requires two AA batteries. Hearing aid compatible. Transmitter and receiver on handset top of airship. Keypad, keys, switches, and on/off hook switch underneath handset on base. Desk set. $60-80.

The Budweiser Beer Bottle Telephone was made in China for Ronsonic Trading Corp. Old Bridge, New Jersey, under license from Anheuser-Busch Incorporated. Model: RON-5230. 1995. 9.5" H x 2.50" diameter. Touch-tone or pulse-dialing switchable. Last-number redial star key. Hearing aid compatible. Transmitter and receiver on back of bottle. Keypad, key, and on/off hook switch underneath bottle. Desk set. $30-50.

The Budweiser Beer Can Telephone was manufactured by TECTEL Incorporated in Taiwan under license from Anheuser-Busch Incorporated. Model: 1071. Circa 1980s. 6.5" H x 2.5" diameter. Touch-tone or pulse-dialing switch. Last-number redial key. Mute key. Transmitter and receiver on front of can. Keypad, keys, switch, and on/off hook switch underneath can. Desk set. $35-65.

The Budweiser Beer Bud Light Can Telephone was made in China for Arrow Trading Co. under license from Anheuser-Busch Incorporated. Model: 5060. Circa 1980s. 6.5" H x 2.5" diameter. Touch-tone dialing. Transmitter and receiver on side of can. Keypad and on/off hook switch underneath can. Desk set. $35-65.

The Burger King Cheeseburger Telephone was manufactured in China by Arrow Trading Co. Incorporated under license from Burger King Corp. 1987. 2.5" H x 4" diameter 8" opened. Touch-tone dialing. Transmitter and receiver inside lid. Keypad and on/off hook switch inside on top of base. Desk set. $70-100.

The Budweiser Beer Bud Light Cup Telephone was made in China for Arrow Trading Co. under license from Anheuser-Busch Incorporated. Model: AR-5043. 1987. 6.75" H x 3.5" diameter. Touch-tone dialing. Last-number redial key. Mute key. Transmitter and receiver on front of cup. Keypad, keys, and on/off hook switch underneath cup. Desk set. $35-55.

The Budweiser Beer Ocean Racer Telephone was made in China for Ronsonic Trading Corp., Old Bridge, New Jersey, under license from Anheuser-Busch Incorporated. Model: RON-135. 1997. 4" H x 3.5" W x 13.5" length. Touch-tone or pulse-dialing switch. Last-number redial key. Flash key for on/off hook. Pause or mute key. Ringer on/off or accelerating boat sound switch. LED indicator light signals incoming calls. Requires two AA batteries. Hearing aid compatible. Transmitter and receiver on handset top half of racer. Keypad, keys, switches, and on/off hook switch underneath handset on base. Desk set. $60-80.

The Burger King Cup Telephone was manufactured in China by Arrow Trading Co. Incorporated under license from Burger King Corp. Model: AR-7500. 1987. 7" H x 3.5" diameter. Touch-tone dialing. Mute key. Last-number redial key. Transmitter and receiver on front of cup. Keypad, keys, and on/off hook switch underneath cup. Desk set. (*Courtesy of Timothy J. Eisenhauer and Jonathan Frank.*) $30-40.

The Busch Cup Telephone was manufactured in China under authority of Anheuser-Busch Incorporated. Model: AR-5042. 1987. 6.75" H x 3.5" diameter. Touch-tone or pulse-dialing switchable. Last-number redial key. Mute key. Transmitter and receiver on front of cup. Keypad, keys, and on/off hook switch underneath cup. Desk set. $35-65.

The Busch Beer Can Telephone was manufactured by ARO Incorporated in Taiwan under license from Anheuser-Busch Incorporated. Model: 1071. Circa 1990s. 6.5" H x 2.5" diameter. Packaged in tall canister as shown. Touch-tone or pulse-dialing switch. Last-number redial key. Mute key. Transmitter and receiver on front of can. Keypad, keys, switch, and on/off hook switch underneath can. Desk set. $35-65.

Above: The Charlie the Tuna Telephone was manufactured in Hong Kong for Star-Kist Foods Incorporated. Model: Charlie the Tuna. 1987. 10" H x 4.5" diameter. Touch-tone or pulse-dialing switch. Last-number redial key. Ringer on/off switch. Transmitter and receiver on back of Charlie. Keypad, key, switches, and on/off hook switch underneath Charlie. Desk set. $60-80.

Top right: The Canada Dry Ginger Ale Can Telephone was manufactured by Tectel Incorporated in Taiwan under license from Anheuser-Busch Incorporated. Circa 1980s. 6.5" H x 2.5" diameter. Packaged in tall canister as shown. Touch-tone or pulse-dialing switch. Last-number redial key. Mute key. Transmitter and receiver on front of can. Keypad, keys, switch, and on/off hook switch underneath can. Desk set. $70-95.

Bottom right: The Coca-Cola Bottle Telephone was precision manufactured in China for Arrow Trading Co. Incorporated, New York, under license from the Coca-Cola Co. Model: 5000. 1983. 9.75" H x 2.25" diameter. Touch-tone or pulse-dialing switchable. Last-number redial key. Mute key. Transmitter and receiver on side of bottle. Keypad, keys, and on/off hook switch underneath bottle. Desk set or wall mount with bracket. $35-65.

The Coca-Cola Coke Big Can Telephone was manufactured by Paul Nelson Industries, Englewood, Colorado. Model: 3030. Circa 1980s. 12.75" H x 8.5" W x 6.5" diameter. Touch-tone dialing. Transmitter and receiver on handset. On/off hook switch on side of can. Handset Stromberg-Carlson. Keypad on top of can. Desk set. (*Courtesy of Betty C.*) $140-180.

The Coca-Cola Coke Can Telephone was manufactured in China under license from the Coca-Cola Co. 1985. 6.25" H x 2.5" diameter. Touch tone. Transmitter and receiver on side of can. Keypad and on/off hook switch underneath can. Desk set. $35-55.

The Coca-Cola Classic Coke Can Telephone was manufactured in China under license from the Coca-Cola Co. Circa 1980s. 6.25" H x 2.5" diameter. Touch tone. Transmitter and receiver on side of can. Keypad and on/off hook switch underneath can. Desk set. $35-65.

The Coca-Cola Coke Disc Telephone was manufactured in China for Telemania Products and Distributed by Kash N' Gold Ltd., Ronkonkoma, New York. Model: HAC Coca-Cola. 1995. 3.25" deep 11.75" diameter. Touch-tone or pulse-dialing switch on side of handset. Ringer music/bell/off switch on back. Bell tone hi/low switch on back. Lights flash/on/off switch on back. Hearing aid compatible. Bell tone or music sounds and front lights up when phone rings. Star and pound keys for various purposes on long distance or other tone/dial systems applications. Last-number redial key. Mute key. Flash key for new on/off hook. Keypad, keys, transmitter, and receiver on handset. On/off hook switch in handset cradle. Desk set or wall mount. $50-75.

The Coca-Cola Diet Coke Can Telephone was manufactured in China under license from the Coca-Cola Co. 1985. 6.25" H x 2.5" diameter. Touch tone. Transmitter and receiver on side of can. Keypad and on/off hook switch underneath can. Desk set. $35-65.

The Coca-Cola Cup with Straw Telephone was precision manufactured in Hong Kong under authority of the Coca-Cola Co. This cup has silver lettering. Model: 6910. Circa 1980s. 6.75" H x 3.5" diameter. Touch-tone or pulse-dialing switchable. Last-number redial key. Mute key. Transmitter and receiver on front of cup. Keypad, keys, and on/off hook switch underneath cup. Desk set. $50-75.

The Coca-Cola Cup with Straw Telephone was precision manufactured in China under authority of the Coca-Cola Co. This cup has white lettering. Model: AR-5040. Circa 1980s. 6.75" H x 3.5" diameter. Touch-tone or pulse-dialing switchable. Last-number redial key. Mute key. Transmitter and receiver on front of cup. Keypad, keys, and on/off hook switch underneath cup. Desk set. $50-75.

The Coca-Cola Olympic Cube Speaker Telephone was manufactured in Hong Kong by Gary Electronics Corp. designed for Communicorp, Atlanta, Georgia under authority of the Coca-Cola Co. Model: HCSP-83. 1994. 4.25" H x 4" W x 4" deep. Touch-tone dialing. Last-number redial key. Speaker phone transmitter and receiver underneath cube. Keypad, key, on/off hook switch, and star key on top of cube. Desk set. $65-95.

The Coca-Cola Desk or Wall Mount Telephone was a premium made for the Coca-Cola Co. Circa 1980s. 1.75" H x 2.5" W x 8" length. Touch-tone dialing. Mute key. Last-number redial key. Ringer on/off switch. Keypad, keys, switch, transmitter, receiver, and on/off hook switch on handset. Desk set or wall mount. $55-75.

The Coca-Cola Polar Bear Telephone was manufactured in China and distributed by Kash N' Gold Ltd., New York, under license from the Coca-Cola Co. Model: HC 900/Tiffany. 1994. 2.5" H x 1.5" W x 7.25" length. Touch-tone or pulse-dialing switch. Last-number redial R/P key. Flash FLH key for new dial tone or call waiting or three-party calling. P-T key for rotary subscribers to change from pulse to tone mode for low-cost long distance and computer services. Keypad, keys, switch, on/off hook switch, transmitter, and receiver on handset. Desk set or wall mount. $40-70.

Left: The Computer Associates Corp. Telephone was made in Taiwan. Circa 1990s. 8.5" H x 4" W x 2.5" deep. Pulse dialing. Transmitter and receiver on front. Keypad and on/off hook switch underneath. Desk set. $15-30.

Right: The Coors Beer Can Telephone was manufactured by Tectel Incorporated in Taiwan under license from Adolph Coors Co. Circa 1980s. 6.25" H x 2.5" diameter. Touch tone. Last-number redial key. Mute key. Transmitter and receiver on front of can. Keypad, keys, and on/off hook switch underneath can. Desk set. $35-65.

The Coors Light Racing Car Telephone was made in China for Columbia TEL-COM Group Incorporated New York, under license from Adolph Coors Co. Model: (HAC) NASCAR-9. 1991. 2" H x 3.5" W x 8.75" length. Touch-tone or pulse-dialing switch. Last-number redial key. Mute key. Keypad, keys, switch, on/off hook switch, transmitter, and receiver underneath car. Desk set. $30-50.

The Crest Toothpaste Man Telephone was manufactured in Taiwan Model: IH-168. Circa 1980s. 10.75" H x 6" diameter. Touch-tone or pulse-dialing switch and ringer on/off switch on top of base. Last-number redial key. Mute key. Transmitter and receiver on front of figure. Keypad, keys, and on/off hook switch underneath base. Desk set. $45-75.

The Doritos Frito Lay Superbowl XIX Football Telephone was made in Taiwan for K-Promotions Inc., Milwaukee, Wisconsin. 1984. 10.5" H x 6.5" W x 8" deep. Touch-tone or pulse-dialing switch. Last-number redial key. Mute key. Keypad, keys, switch, on/off hook switch, transmitter, and receiver on handset. Desk set. $55-80.

Below: The Gasoline Pump Telephones were manufactured in Taiwan by Synanon. 1984. 9" H x 3.25" diameter. Touch-tone dialing. Last-number redial key. Mute key. Transmitter and receiver on front of pump. Keypad, keys, and on/off hook switch underneath pump. Desk set. L to R Good Gulf, Texaco, Shell, Sunoco, Conoco, and Union 76. $65-90.

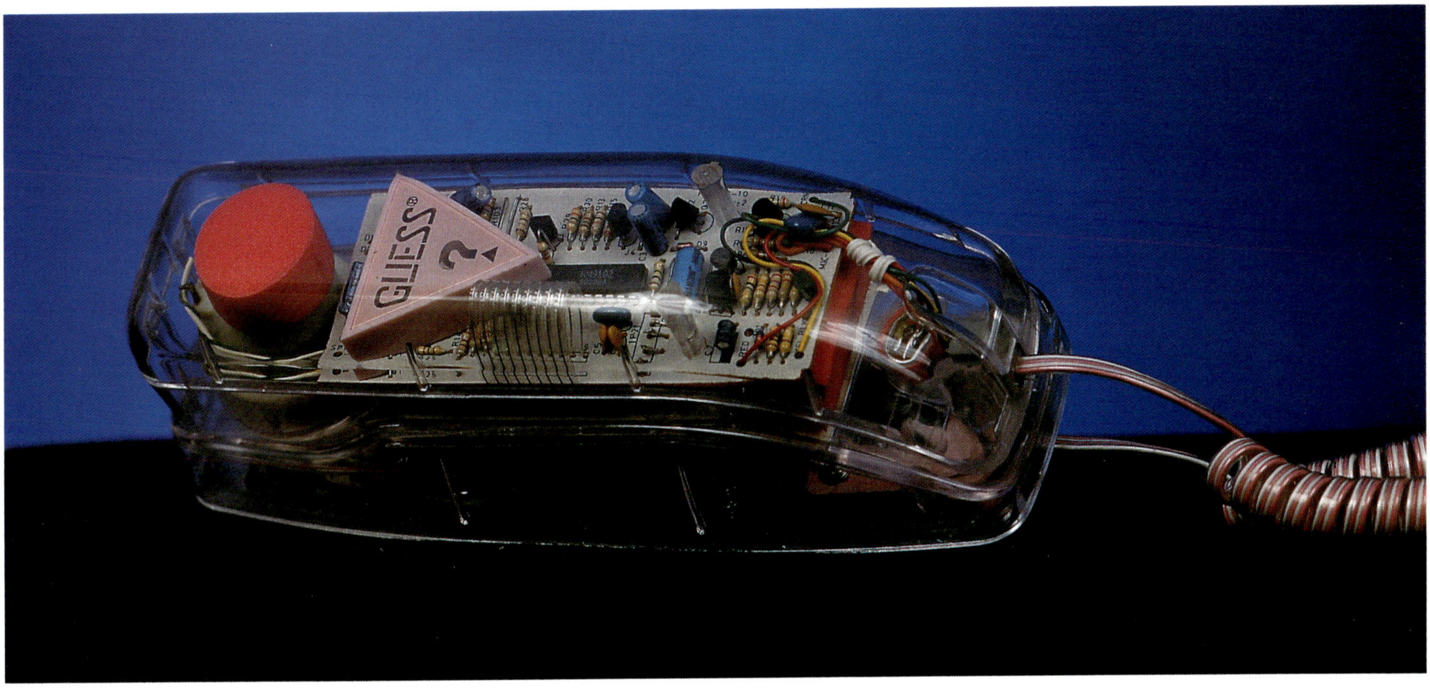

The Guess ? Products Telephone was made in China for Wholewell Industrial Ltd. Model: E400. Circa 1980s. 3" H x 2.75" W x 7.75" length. Touch-tone or pulse-dialing switch. Last-number redial key. Keypad, key, switch, transmitter, and receiver on handset. Ringer on/off switch, light on/off switch, and on/off hook switch on base. Desk set. $55-75.

The Guess ? Products with Alarm Clock Telephone was made in China for Wholewell Industrial Ltd. Model: E306. Circa 1980s. 2.75" H x 8" W x 6" deep. Touch-tone or pulse-dialing switch on back of base. Ringer on/off switch on back of base. Hearing aid compatible. Clock controls underneath base. Transmitter and receiver on handset. Keypad and on/off hook switch on base. Desk set. $55-75.

HARLEY-DAVIDSON®

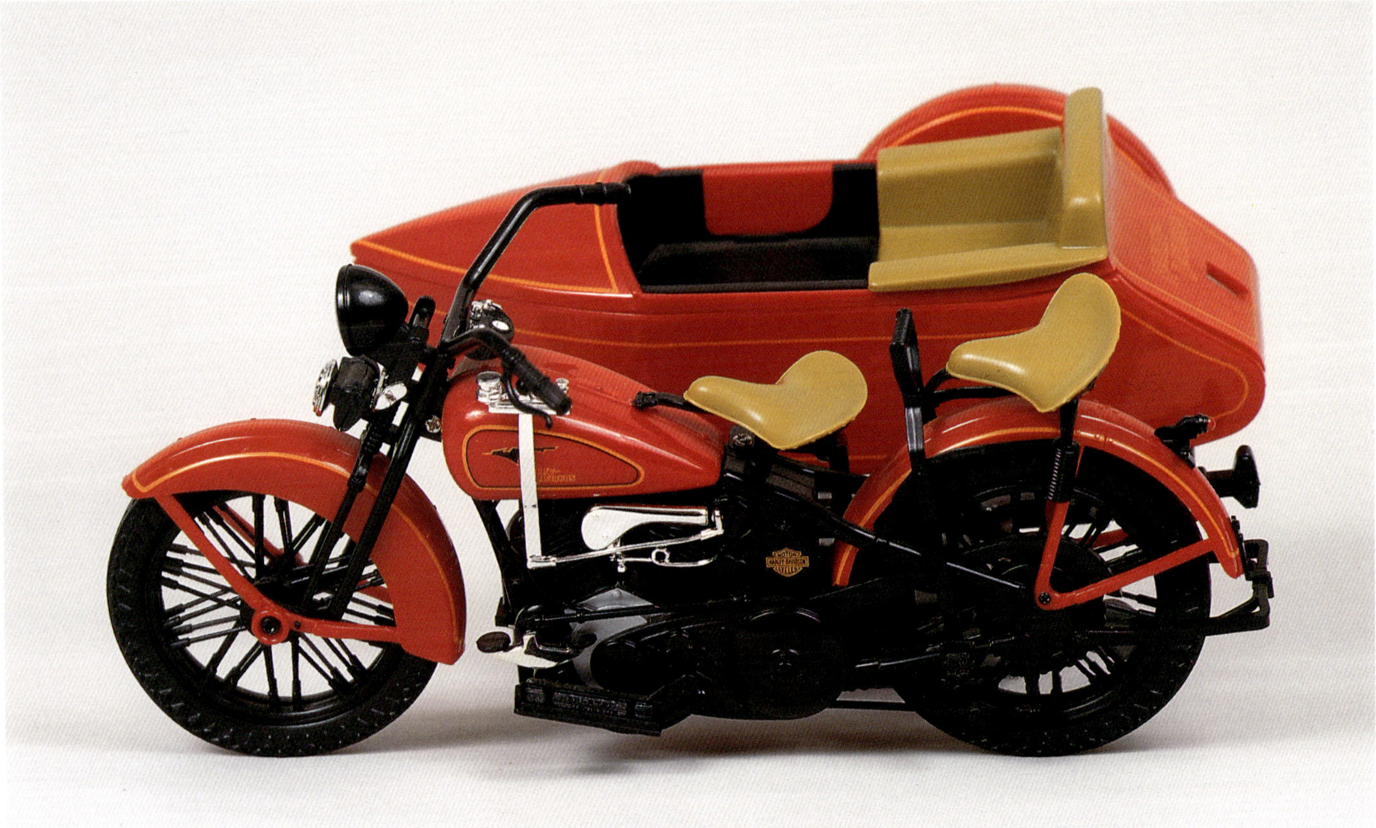

Opposite page:
Top: The Harley-Davidson Motorcycle Telephone was manufactured in China for Telemania Products and distributed by Kash N' Gold Ltd., Ronkonkoma, New York. Model: Harley-Davidson Telephone. 1993. 9" H x 5.5" W x 15" length. Touch-tone or pulse-dialing switch underneath seat handset. Ringer on/off switch underneath motorcycle. Hearing aid compatible. Horn sounds and headlight flashes when phone rings. Star and pound keys for various purposes on long distance or other tone/dial systems applications. Last-number redial key. Mute key. Keypad, keys and on/off hook switch underneath seat handset. Transmitter and receiver on seat handset. Desk set. $55-75.
Bottom: The Harley-Davidson Motorcycle With Sidecar Telephone was manufactured in China for Telemania Products and distributed by Kash N' Gold Ltd., Ronkonkoma, New York. Model: Harley-Davidson Sidecar. 1997. 5.5" H x 10" W x 12" deep. Touch-tone or pulse-dialing switch. Last-number redial key. Ringer or sound switch underneath motorcycle. Keypad, key, and on/off hook switch underneath sidecar seat handset. Transmitter and receiver on sidecar seat handset. Desk set. (*Courtesy of Betty C.*) $60-70.

The Heinz 57 Tomato Catsup Bottle Telephone was made in Taiwan for H.J. Heinz Co., Pittsburgh, Pennsylvania. Circa 1980s. 8.5" H x 2.5" diameter. Touch-tone dialing. Last-number redial key. Mute key. Keypad, keys, and on/off hook switch underneath bottle. Transmitter and receiver on front of bottle. Desk set. $35-65

Below: The Hertz "The #1 Way to Rent a Car" Telephone was made in Taiwan for the Hertz Corp. Circa 1980s. 8" H x 4" W x 2.5" deep. Touch-tone dialing. Last-number redial key. Mute key. Keypad, keys, and on/off hook switch underneath base. Transmitter and receiver on front. Desk set. $65-95.

The Hershey's Chocolate Milk Container Telephone was made in Taiwan for Arrow Trading Co., Incorporated, New York, under license from the Hershey's Foods Corp. Model: 6000. 1985. 9" H x 2.75" W x 2.75" deep. Touch-tone dialing. Last-number redial key. Mute key. Keypad, keys and on/off hook switch underneath container. Transmitter and receiver on front of container. Desk set. $95-140.

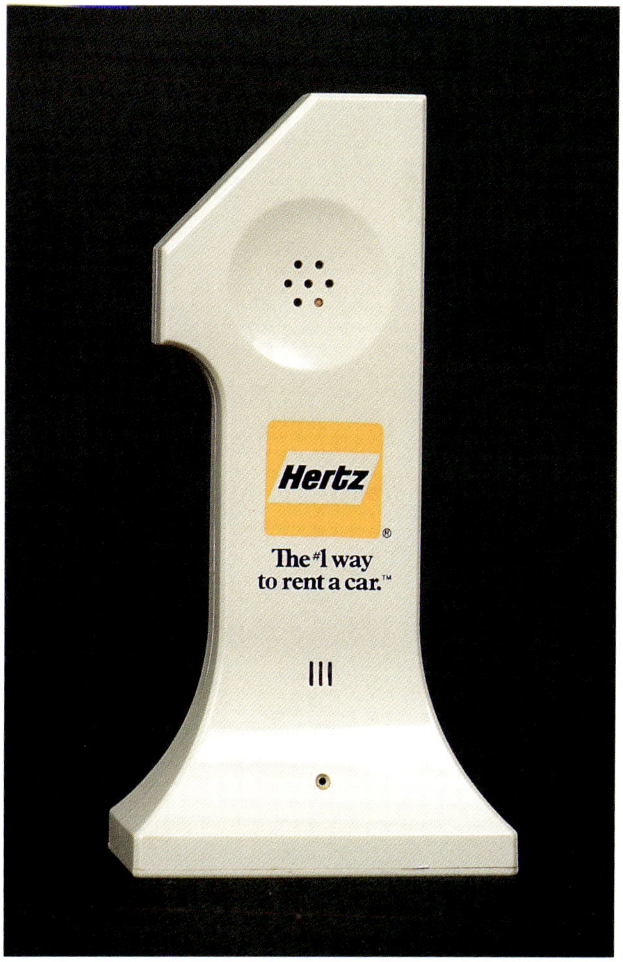

The Hawaiian Punch 'Punchy' Telephone was made in China for the Hawaiian Punch Co. Circa 1980s. 10.5" H x 9" W x 6" deep. Touch-tone or pulse-dialing switch under base. Ringer on/off switch under base. On/off hook switch in cradle below handset. Transmitter and receiver on handset. Keypad on top of base. Desk set. $140-180.

Above: The Jeff Gordon Race Rig Fone DuPont Semi-Truck Telephone was made in China for Columbia TEL-COM Group Inc., New York, under license from DuPont Corp. Model: Truck-24. 1997. 3.75" H x 2.25" W x 14" length. Touch-tone or pulse-dialing switch. Last-number redial key. Mute key. Keypad, keys, switch, on/off hook switch, transmitter, and receiver underneath truck. Truck has room to carry two racing cars, extra engines, and all the team's equipment. Authentically styled Kenworth tractor pulls the rig. Desk set. $30-45.

Below: The Jolly Green Giant Telephone was made in Hong Kong for the Pillsbury Co. Model: C 168. 1984. 13.5" H x 9.75" W x 5.25" deep. Touch-tone dialing. Last-number redial key. Mute key. Ringer on/off switch. Keypad, keys, switch, on/off hook switch, transmitter, and receiver on handset. Desk set. $65-85.

Left: The Jim Beam Whiskey 200-Years Anniversary Telephone was made in China for the Jim Beam Corp. Circa 1990s. 19.75" H x 3" H x 3" deep. Touch-tone or pulse-dialing switch. Last-number redial key. Mute key. Keypad, keys, switch, and on/off hook switch underneath bottle. Transmitter and receiver on back of bottle. Desk set. (*Courtesy of Betty C.*) $35-45.

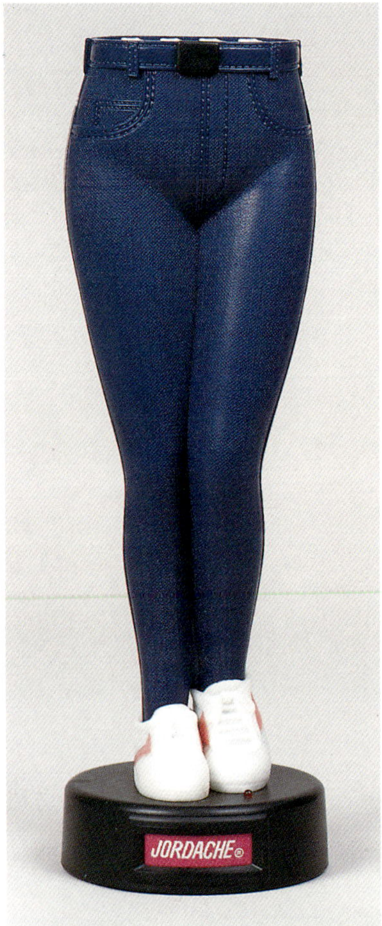

The Jordache Jeans Telephone was made in China for the Jordache Jeans Co. Model: HAC RON-190. Circa 1980s. 10.5" H x 3.5" diameter. Touch-tone dialing. Last-number redial key. Mute key. Ringer hi/low/off switch underneath base. Keypad and keys on top of jeans. Transmitter and receiver on back of jeans. On/off hook switch underneath base. Desk set. $75-95.

The Keebler telELFphone Telephone was made in Hong Kong for the Keebler Co. 1985. 13" H x 10" W x 6.75" deep. Touch-tone and pulse-dialing switch. Mute key. Last-number redial key. Keypad, keys, switch, on/off hook switch, transmitter, and receiver on handset. Desk set. $75-95.

The Kawasaki Jet Ski Telephone was manufactured in China for Telemania distributed by Kash N' Gold. Licensed By Kawasaki Motors Corp., U.S.A. Model: Jet Ski Telephone. 1993. 5.5" H x 4.75" W x 11.75" length. Touch-tone or pulse-dialing switch. Last-number redial key. Beep-beep horn sounds when phone rings. Transmitter and receiver located on seat handset. Keypad, on/off hook switch, key, and switch underneath seat handset. Desk set. $55-75.

The Lego Super Blocks Telephone was manufactured in Taiwan for Tyco Industries Incorporated, Moorestown, New Jersey. 1983. 4.5" H x 7.5" W x 7.5" deep. Touch-tone dialing. Mute key. Last-number redial key. Slide out drawer on side holds extra blocks. Keypad and keys on top of base. Transmitter and receiver on handset. On/off hook switch on cradle under handset. Desk set. $65-95.

The LA Gear Tennis Shoe Telephone was made in China for Custom Phones Incorporated, Indianapolis, Indiana. Model: TENNY-TS300. Circa 1990s. 13" H x 10" W x 7.5" deep. Touch-tone or pulse-dialing switch. Last-number redial key. Ringer on/off switch. Keypad, key, switches, on/off hook switch, transmitter, and receiver underneath shoe. Desk set. $40-60.

The McDonald's Racing Car Telephone was made in China for Columbia TEL-COM Group Incorporated, New York, under license from McDonald's Corp. Model: (HAC) McDonald-27. 1993. 2" H x 3.5" W x 8.75" length. Touch-tone or pulse-dialing switch. Last-number redial key. Mute key. Keypad, keys, switch, on/off hook switch, transmitter, and receiver underneath car. Desk set. $30-50.

The McDonald's Ronald Sitting Telephone was manufactured in Canada by Rubicam Creative Products Ltd., Toronto, Canada. Under license of the McDonald's Corp. Model: 6003-0001. 1980. 14.25" H x 12" W x 12" deep. Touch-tone dialing. Ringer on/off switch on top of base. Ringer volume control switch underneath base. Keypad on front of base. On/off hook switch, transmitter, and receiver on handset. Desk set. $400-600.

The McDonald's French Fries Telephone was made in China for World Wide Design under license from McDonald's Corp. Model: FF-389. 1993. 6.75" H x 4.25" W x 2" deep. Touch-tone or pulse-dialing switch. Last-number redial key. Mute key. Hearing aid compatible. Keypad, keys, switch, on/off hook switch, transmitter, and receiver on back. Button on side opens back like flip phone. Desk set. $50-75.

The McDonald's Ronald Standing Telephone was manufactured in Hong Kong by Merton Co. Ltd. for H. P. International. Distributed by Group II Communications Incorporated, Milwaukee, Wisconsin. Model: 729ZT. 1985. 10.5" H x 3.75" W x 2.5" deep. Touch-tone dialing. Keypad, transmitter, and receiver on back. On/off hook switch underneath base. Desk set. $100-150.

The Miller Genuine Draft Beer Can Telephone was manufactured by TECTEL Inc. in Taiwan under license from Miller Brewing Co. Model: RON-6000. 1994. 6.5" H x 2.5" diameter. Touch-tone or pulse-dialing switchable. Last-number redial pound key. Mute star key. Transmitter and receiver on front of can. Keypad, keys, and on/off hook switch underneath can. Desk set. $45-65.

The Miller Lite Beer Bottle Telephone was made in China for Ronsonic Trading Corp., Old Bridge, New Jersey, under license of the Miller Brewing Co. Model: RON-6300. 1994. 9.5" H x 2.25" diameter. Touch-tone dialing. Last-number redial key. Keypad, key, and on/off hook switch underneath bottle. Receiver and transmitter on backside of bottle. Desk set or wall mount with bracket. $35-65.

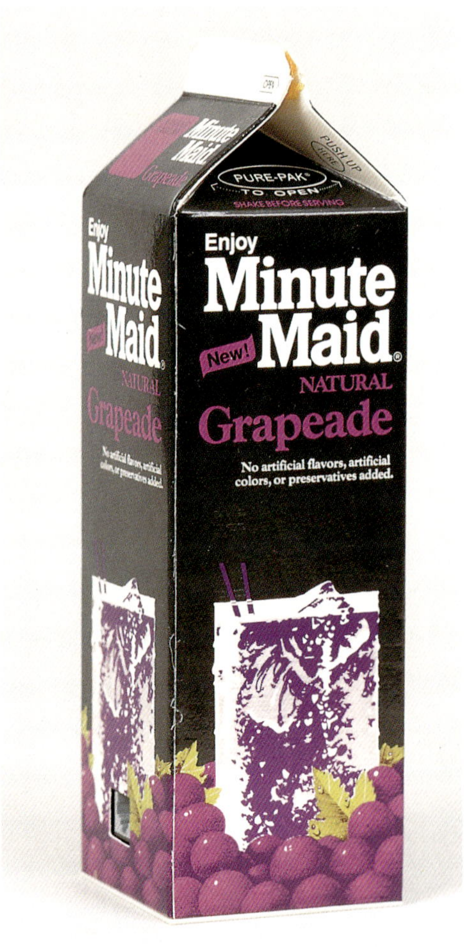

The Minute Maid Grapeade Container Telephone was made in Taiwan for Arrow Trading Co. Inc., New York, under license from Coca-Cola Foods. Model: 6000. 1985. 9.25" H x 2.75" W x 2.75" deep. Touch-tone dialing. Last-number redial key. Mute key. Keypad, keys, and on/off hook switch underneath container. Transmitter and receiver on front of container. Desk set or wall mount with bracket. $100-140.

The NFL Raiders Football Telephone was made in China for Columbia TEL-COM Corp. Logo was available for all NFL teams. Model: NFL-28. Circa 1990s. 9.5" H x 6.5" W x 8.5" deep. Touch-tone or pulse-dialing switch. Last-number redial key. Mute key. Keypad, keys, switch, on/off hook switch, receiver, and transmitter on handset. Desk set. $40-60.

The Oklahoma Sooners Desk Set Telephone was made for Western Electric Co. Circa 1980s. 3" H x 6.5" W x 8.5" deep. Touch-tone dialing. Keypad and on/off hook switch on top of base. Transmitter and receiver on handset. Desk set. $45-70.

The Oklahoma Sooners Recruiting Machine Car Telephone was made in Hong Kong for Dialfone Ltd. 1986. 2" H x 3.25" W x 7.5" length. Touch-tone dialing. Last-number redial key. Mute key. Ringer on/off switch. Keypad, keys, switch, on/off hook switch, transmitter, and receiver underneath car. Desk set. $45-70.

The Oklahoma State Recruiting Machine Car Telephone was made in Hong Kong. Model: RON-240. Circa 1980s. 2" H x 3.25" W x 7.5" length. Touch-tone dialing. Last-number redial key. Mute key. Ringer on/off switch. Keypad, keys, switch, on/off hook switch, transmitter, and receiver underneath car. Desk set. $45-70.

The Pepsi Candlestick Telephone was made in the U.S.A. by American Telecommunications Corp., El Monte, California. 1974. 12" H x 5.75" diameter. Pulse dialing. Transmitter mouthpiece. Receiver handset. On/off hook switch on side of stick. Desk set. $95-140.

The Pepsi Vending Machine Telephone was made in China for PepsiCo Incorporated by PF Sound Collection, Stanford, Connecticut. 1995. 13.75" H x 8" W x 2" deep. Touch-tone or pulse-dialing switch. Ringer on/off switch. In-use LED indicator. Keypad, switches, LED, on/off hook switch, transmitter, and receiver on handset. Desk set or wall mount. $70-90.

The Pepsi-Cola Can Telephone was manufactured by Tectel Incorporated in Taiwan used with permission of PepsiCo Incorporated. Model: 1073. Circa 1980s. 6.5" H x 2.5" diameter. Packaged in tall canister as shown. Touch-tone or pulse-dialing switchable. Last-number redial key. Mute key. Transmitter and receiver on front of can. Keypad, keys, and on/off hook switch underneath can. Desk set. $95-140.

The Ore-Ida French Fry Telephone was made in China for World Wide Design under license from Ore-Ida Foods. Model: FF-389. 1991. 6.75" H x 4.25" W x 2" deep. Touch-tone or pulse-dialing switch. Last-number redial pound key. Mute star key. Keypad, keys, switch, on/off hook switch, transmitter, and receiver on back. Button on side opens back like flip phone. Desk set. $70-95.

The Pizza Inn Man Telephone was manufactured in Taiwan Circa 1980s. 10.5" H x 5" diameter. Touch-tone dialing. Last-number redial key. Mute key. Keypad, keys, and on/off hook switch underneath base. Transmitter and receiver on front of figure. Desk set. $35-70.

The Pillsbury Dough Boy Telephone was made in Hong Kong for the Pillsbury Co. Model: 888. 1984. 15" H x 8.5" W x 7" deep. Touch-tone or pulse-dialing switch on front of base. Mute key. Last-number redial key. On/off hook switch on arm. Keypad, keys, transmitter, and receiver on handset. Desk set. $450-600.

The Quantas Airlines Koala Bear Telephone was made in Taiwan by Exicom under license of Quantas Airlines. Model: KLA -1. Circa 1990s. 18" H. Touch-tone dialing. Keypad, on/off hook switch, transmitter, and receiver on handset. Desk set. (*Courtesy of Timothy J. Eisenhauer and Jonathan Frank.*) $100-125.

The Red Dog Bottle Cap Telephone was manufactured in China by Longwood Industries for Telemania Products and distributed by Kash N' Gold Ltd., Ronkonkoma, New York. 1996. 3.25" deep 12.5" diameter. Touch-tone or pulse-dialing switch on side of handset. Ringer music/bell/off switch on back. Bell tone hi/low switch on back. Lights flash/on/off switch on back. Hearing aid compatible. Bell tone or music sounds and front lights up when phone rings. Star and pound keys for various purposes on long distance or other tone/dial systems applications. Last-number redial key. Mute key. Flash key for new on/off hook. Keypad, keys, transmitter and receiver on handset. On/off hook switch in handset cradle. Desk set or wall mount. $60-80.

The Raid Bug Telephone was manufactured for the Raid Corp. Circa 1980s. 8.5" H x 5" W x 4.25" deep. Touch-tone dialing. Last-number redial key. Mute key. Ringer on/off switch. Keypad, keys, switch, and on/off hook switch underneath base. Transmitter on front of barrel. Receiver in front on top of base. Desk set. $125-160.

The Seagram's Whiskey Bottle Telephone was made in China for Custom Phones Incorporated, Indianapolis, Indiana. Model: BOTTL. Circa 1990s. 13" H x 10" W x 7.5" deep. Touch-tone or pulse-dialing switch. Last-number redial key. Mute key. Keypad, keys, switch, on/off hook switch, transmitter, and receiver on handset. Desk set. $75-100.

The 7-Up Can Telephone was manufactured by Tectel Incorporated in Taiwan Circa 1980s. 6.25" H x 2.5" diameter. Touch tone. Last-number redial key. Mute key. Transmitter and receiver on front of can. Keypad, keys, and on/off hook switch underneath can. Desk set. $55-75.

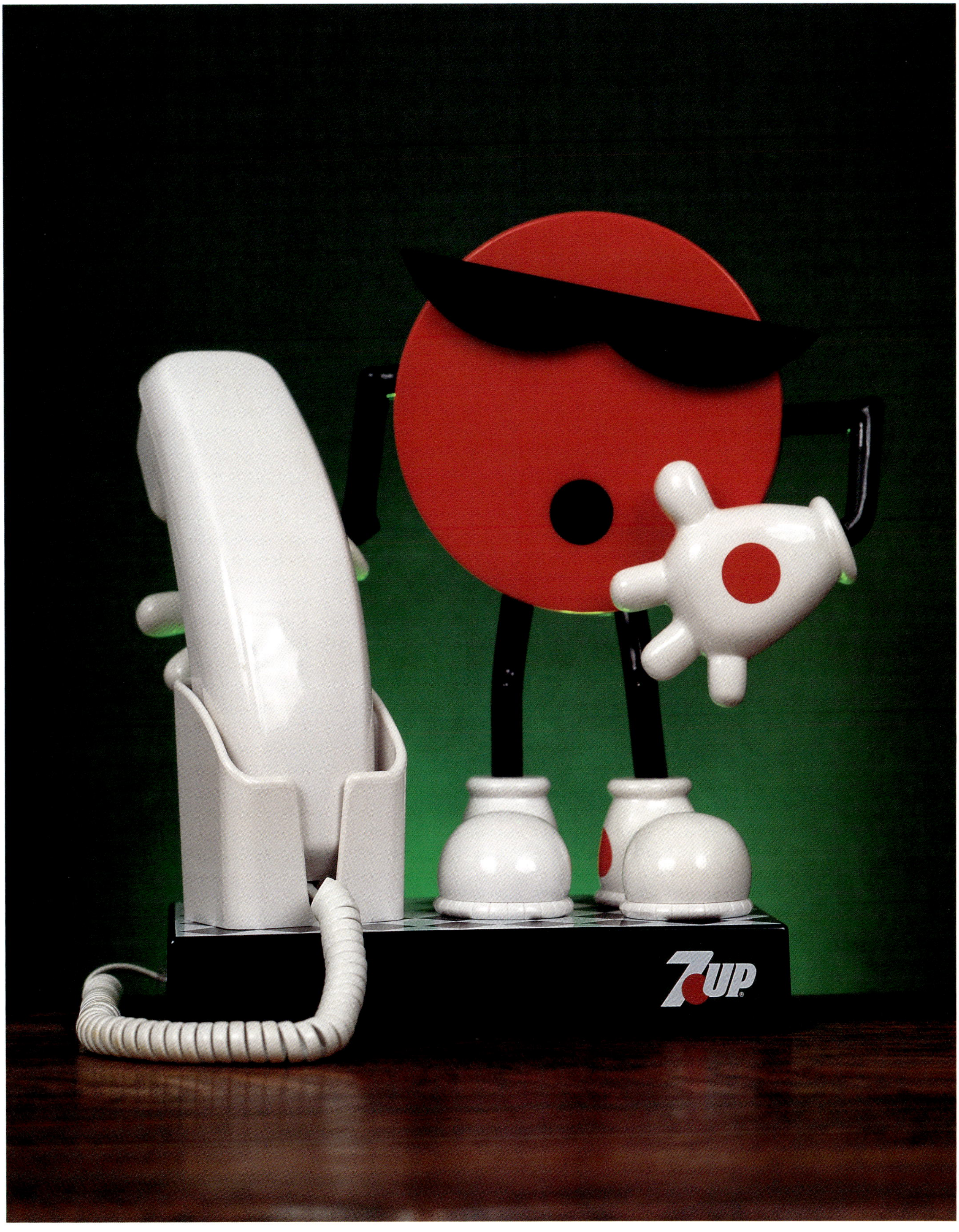

The 7-Up Spot Telephone was manufactured in China. 1990. 11.5" H x 11" W x 5" deep. Touch-tone or pulse-dialing switch. Last-number redial key. Reset key for dial tone. Keypad, keys, switch, transmitter, and receiver on handset. On/off hook switch on top of base. Desk set. $85-130.

The Skittles Bite-size Candies Telephone was manufactured in China. Model: CP336. Circa 1980s. 2.75" H x 4.75" diameter. Touch-tone dialing. Keypad, transmitter, receiver, and on/off hook switch underneath lid. Desk set. $70-110.

The Stanley Tape 30-foot Powerlock Telephone was made in Hong Kong for Premium Resources International under authority of Stanley Works. Model: 33-430. Circa 1990s. 1.75" H x 3.25" W x 3.25" deep. Touch-tone dialing. Last-number redial pound key. Mute star key. Keypad, keys, switch, on/off hook switch, transmitter, and receiver inside tape. Desk set. $60-80.

The Call for Smirnoff Vodka Telephone was manufactured in Hong Kong. Model: HAC TP-788. Circa 1980s. 14.5" H x 6.5" diameter. Touch-tone dialing. Ringer on/off switch on handset. Keypad, transmitter, receiver, and on/off hook switch on handset. Desk set. $70-95.

The Sports Illustrated Football Telephone was made in Taiwan Model: TK-010. Circa 1980s. 6" H x 3.25" W x 4.5" deep. Touch-tone dialing. Last-number redial key. Mute key. Ringer on/off switch. Keypad, keys, switch, on/off hook switch, transmitter, and receiver under lid. Desk set. $70-95.

The Sports Illustrated Tennis Shoe Telephone was made in China. Model: 0202. Circa 1980s. 4" H x 3.75" W x 10" length. Touch-tone dialing. Last-number redial key. Mute key. Ringer on/off switch. Keypad, keys, switch, on/off hook switch, transmitter, and receiver underneath shoe. Desk set. $70-95.

The Swatch Twin Telephone was manufactured in Taiwan Model: NXR 243. Circa 1990s. 2.73" H x 2.75" W x 9.5" length. Both handset and base have transmitter and receiver for two people to converse at the same time. Touch-tone or pulse-dialing switch on side. Last-number redial key. Mute key. Keypad on top of base. On/off hook switch on top and on base. Desk set. $35-60.

Far left: The Tang Mouth Telephone was made in Taiwan for General Foods Inc. Model: M5W 1H1. Circa 1990s. 8.75" H x 5.5" W x 5" diameter. Touch-tone or pulse-dialing switch. Ringer on/off switch. Receiver is Tang mouth. Transmitter on top of base in front. Keypad, switches, and on/off hook switch underneath base. Desk set. $95-125.

Left: The Tetley Tea Telephone was manufactured in Taiwan for the Tetley Co. Circa 1990s. 9.75" H x 5.25" diameter. Touch-tone or pulse-dialing switch. Last-number redial key. Pause or mute key. Transmitter in front on top of base. Receiver is Tetley's mouth. Keypad, keys, switch, and on/off hook switch underneath. Desk set. $125-150.

The Ten-K Potassium Chloride Telephone was made in Taiwan Circa 1980s. 2.75" H x 3" W x 7" length. Touch-tone dialing. Ringer on/off switch. Keypad, switch, on/off hook switch, transmitter, and receiver underneath. Desk set. $75-95.

Below: The Terra Seed Corn Telephone was made in China for the Terra Seed Corp. Model: CP-8000. Circa 1980s. 2.25" H x 3" W x 10.25" length. Touch-tone or pulse-dialing switch. Last-number redial key. Mute key. Keypad, keys, on/off hook switch, transmitter, and receiver underneath corn. Desk set. $20-40.

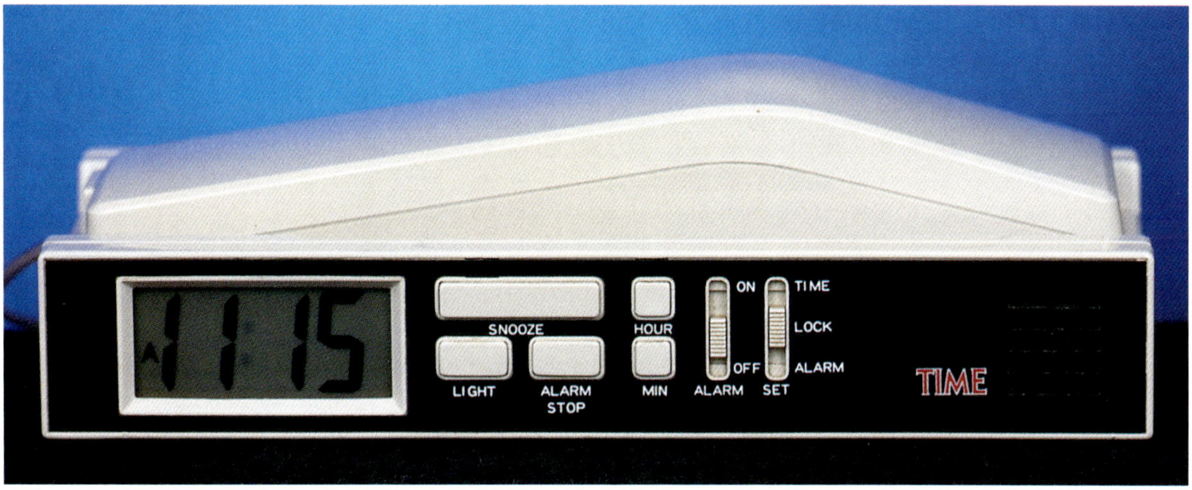

The Time Magazine Telephone was made in Taiwan. Model: TK-007. Circa 1990s. 2.25" H x 4" W x 8.25" length. Touch-tone dialing. Last-number redial key. Mute key. Ringer on/off switch. Battery operated clock in base. Requires one AAA battery. Keypad, keys, switch, on/off hook switch, transmitter, and receiver on handset. Desk set. $15-30.

The TV Guide Telephone was made in Taiwan. Circa 1990s. 8.5" H x 4" W x 2.5" deep. Pulse dialing. Transmitter and receiver on front. Keypad and on/off hook switch underneath. Desk set. $15-30.

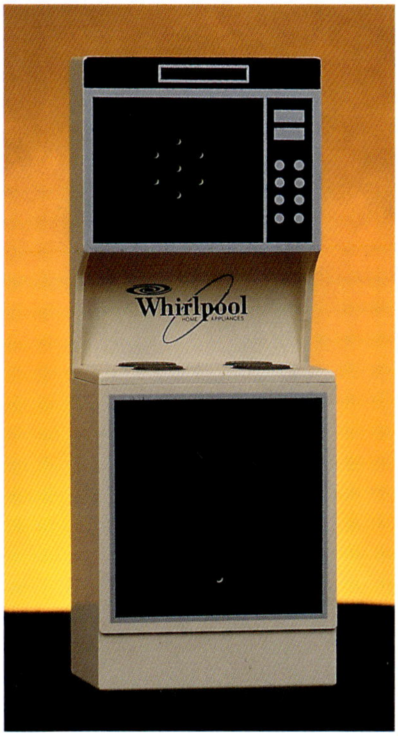

The Whirlpool Home Appliances Telephone was made in Taiwan by Synanon. Model: GP-865. 1985. 8.5" H x 3.25" W x 2.25" deep. Touch-tone dialing. Last-number redial key. Mute key. Keypad, keys, and on/off hook switch underneath appliance. Transmitter and receiver on front of appliance. Desk set. $95-140.

The ZEP Telephone was manufactured by Tectel Incorporated in Taiwan for Zep Manufacturing Co. Circa 1990s. 6.25" H x 2.5" diameter. Packaged in tall canister as shown. Touch tone. Last-number redial key. Mute key. Transmitter and receiver on front of can. Keypad, keys, and on/off hook switch underneath can. Desk set. (*Courtesy of Timothy J. Eisenhauer and Jonathan Frank.*) $15-30.

Chapter III

GREAT SHAPES

Figural Telephones and Etcetera

The popularity of figural telephones has mushroomed in the 1990s and every conceivable shape is being manufactured. These real working telephones are found in the forms of objects such as food, animals, sports equipment, fish, automobiles, trucks, planes, trains, and human parts.

The Etcetera includes all those novelty phones that do not fit neatly into other categories.

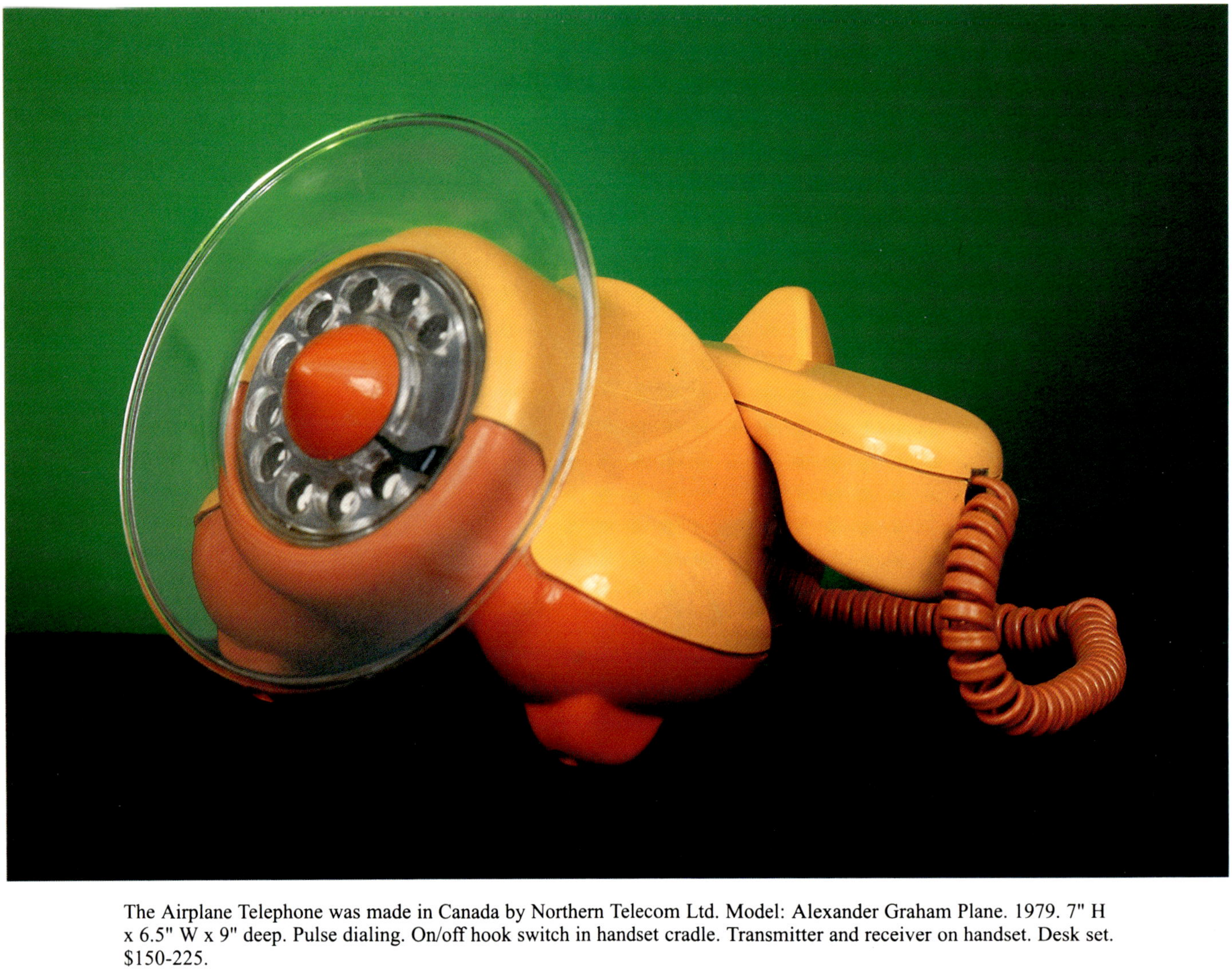

The Airplane Telephone was made in Canada by Northern Telecom Ltd. Model: Alexander Graham Plane. 1979. 7" H x 6.5" W x 9" deep. Pulse dialing. On/off hook switch in handset cradle. Transmitter and receiver on handset. Desk set. $150-225.

Left: The Apple Telephone was made in China for Ronsonic Trading Corp., Old Bridge, New Jersey. Model: RON-65 Apple Phone. 1985. 5" H x 4" diameter. Touch-tone or pulse-dialing switch. Keypad, key, switch, on/off hook switch, transmitter, and receiver underneath lid. Desk set. (*Courtesy of Betty C.*) $30-40.

Center: The Barking Hound or Dalmatian Telephone was made in China for Telemania Products and distributed by Kash N' Gold Ltd., Ronkonkoma, New York. Model: DG 555. 1996. 3" H x 4" W x 8" length. Touch-tone or pulse-dialing switch. Last-number redial key. Keypad, switch, key, on/off hook switch, transmitter, and receiver located underneath dog. Desk set. (*Courtesy of Betty C.*) $30-40.

Bottom: The Banana Telephone was made in China. 1992. 2.5" H x 2.5" W x 10.75" length. Touch-tone or pulse-dialing switch. Last-number redial key. Mute key. Keypad, switch, keys, on/off hook switch, transmitter, and receiver underneath banana. Desk set. (*Courtesy of Betty C.*) $30-40.

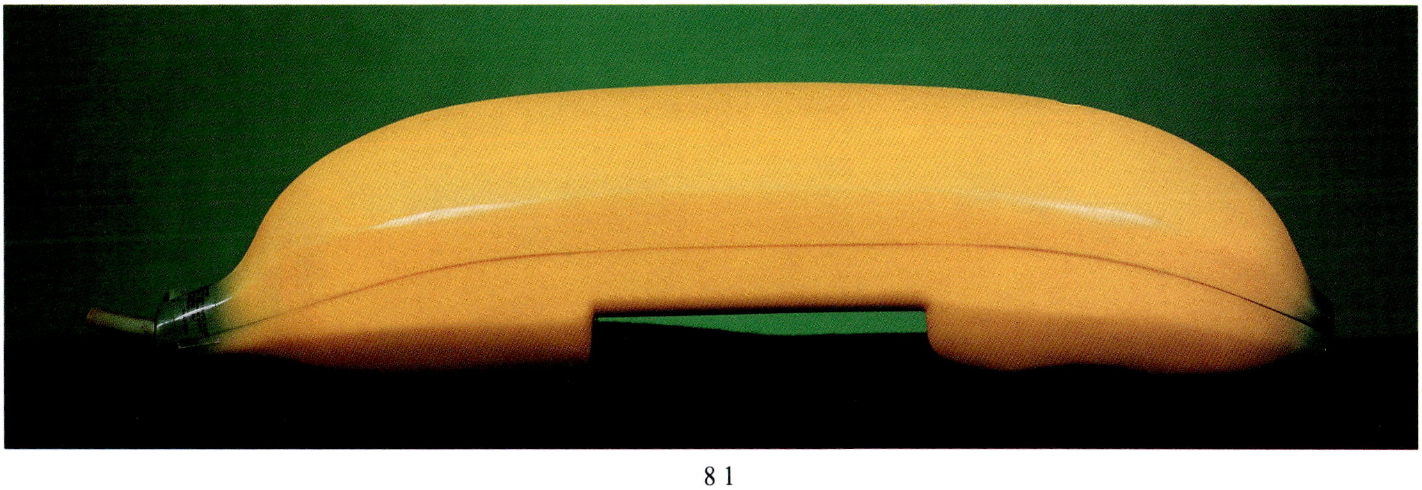

The Basketball Fone Telephone was made in China. Model: BK-391. Circa 1990s. 5" H x 4.5" diameter. Touch-tone dialing. Last-number redial key. Mute key. Keypad, keys, on/off hook switch, transmitter, and receiver underneath lid. Desk set. (*Courtesy of Betty Chapman.*) $30-40.

The Blimp Telephone was made in China for Ronsonic Trading Corp., Old Bridge, New Jersey. Model: RON-109. Circa 1990s. 5" H x 3.75" W x 10.5" length. Touch-tone or pulse-dialing switch. Last-number redial key. Pause or mute key. Ringer on/off switch. Flash key for new on/off hook. Hearing aid compatible. Transmitter and receiver on handset. Keypad, keys, switches, and on/off hook switch in handset cradle. Desk set. (*Courtesy of Timothy J. Eisenhauer and Jonathan Frank.*) $20-30.

The Butterfly Telephone was made in China for Ronsonic Trading Corp., Old Bridge, New Jersey. Model: RON-130. Circa 1990s. 4" H x 10.5" W x 11" length. Touch-tone or pulse-dialing switch. Hearing aid compatible. Last-number redial key. Reset key for new dial tone. Wings light up when phone rings. Keypad, keys, switch, on/off hook switch, transmitter, and receiver on handset. Desk set. (*Courtesy of Betty C.*) $65-75.

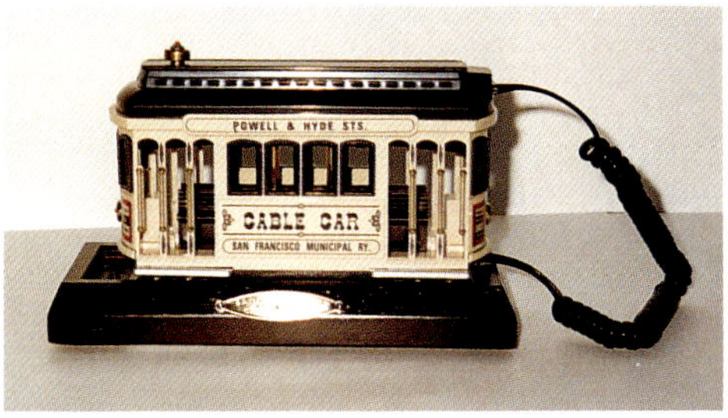

The Cable Car Telephone was made in China for Polyflame Concepts Incorporated. Circa 1990s. 6" H x 9" W. Touch-tone or pulse-dialing switch. Ringer on/off switch. Keypad, keys, switch, on/off hook switch, transmitter, and receiver on handset. Desk set. (*Courtesy of Timothy J. Eisenhauer and Jonathan Frank.*) $30-50.

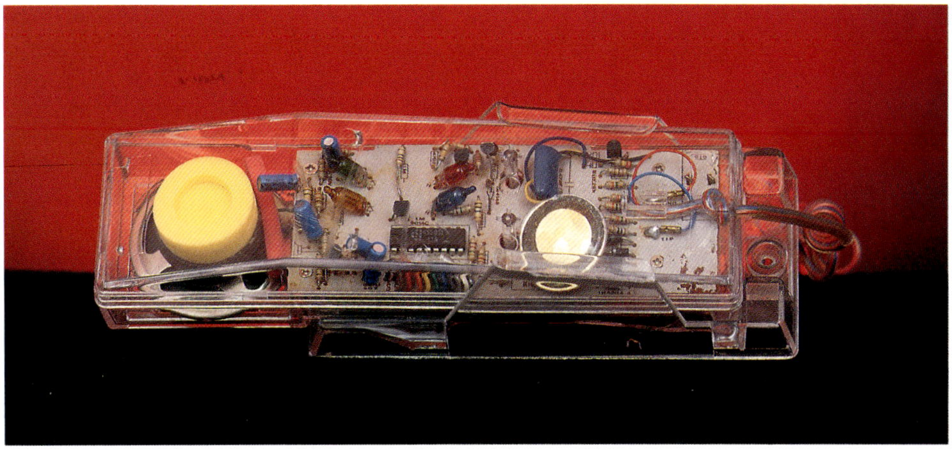

The Clear Desk Set or Wall Mount Telephone. Model: HAC 228HA. Circa 1990s. 1.75" H x 2.5" W x 8.5" length. Touch-tone dialing. Last-number redial key. Mute key. Ringer on/off switch. Keypad, keys, switch, on/off hook switch, transmitter, and receiver on handset. Desk set or wall mount. $10-20.

The Cat Telephone was made in China for Telemania Products and distributed by Kash N' Gold Ltd., Ronkonkoma, New York. Model: Cat Phone. Circa 1990s. 5.5" H x 5.25" W x 11.75" length. Touch-tone or pulse-dialing switch. Last-number redial key. Keypad, keys, on/off hook switch, transmitter, and receiver underneath cat. Desk set. (*Courtesy of Betty C.*) $30-40.

Above: The Clear See Thru with Bell Telephone was made in U.S.A. for Teleconcepts Inc., West Hartford, Connecticut. Model: 460041. 1984. 4.5" H x 8" W x 8.25" deep. Touch-tone or pulse-dialing switch on back of base. Keypad, on/off hook switch, transmitter, and receiver on top of base. Desk set. $20-30.

Left: The Clear Plastic Big Cylinder Telephone. Circa 1990s. 17.5" H x 5.25" diameter. Rotary dialing. On/off hook switch on side marked Telemania. Transmitter and receiver on handset. Handset marked ITT. Desk set. (*Courtesy of Betty C.*) $100-125.

Below: The Clear Desk Set Telephone was made by Cortelco of Corinth, Mississippi, for the ITT Corp. Model: 2500. 1991. 4.25" H x 8" W x 8.75" deep. Touch-tone dialing. On/off hook switch in handset cradle. Transmitter and receiver on handset. Desk set. $60-70.

The Clear See Thru Plastic Trimline Telephone was made by Unisonic. Circa 1990s. 3.75" H x 3" W x 8.5" length. Touch-tone or pulse-dialing switch. Last-number redial key. Flash or mute key. Ringer on/off switch on base. Keypad, keys, switch, transmitter, and receiver on handset. On/off hook switch in base cradle. Desk set or wall mount. $15-20.

The Computer CP-2000 Telephone was made in Taiwan. Model: CT4001. 1985. 4.25" H x 2.75" W x 4" deep. Touch-tone dialing. Transmitter and receiver on front. Keypad and on/off hook switch located underneath computer. Desk set. (*Courtesy of Betty C.*) 60-70.

The Cow Telephone was made in China. Model: HAC-CS-323. 1996. 4.75" H x 5" W x 9" length. Touch-tone or pulse-dialing switch. Last-number redial key. Cow moo sound or ringer switch. Keypad, switches, key, on/off hook switch, transmitter, and receiver underneath cow. Desk set. (*Courtesy of Betty C.*) $35-45.

The Diner '50s Juke Box Telephone was made in China for Polyconcept USA Incorporated. Circa 1990s. 12" H x 11.5" W x 6.5" deep. Touch-tone dialing. Last-number redial key. Ringer and volume controls. Replaceable index cards. Ringer indicator on top. Transmitter and receiver located on handset. Keypad and on/off hook switch on handset. Wall set. (*Courtesy of Betty C.*) $75-85.

The Dinosaur Telephone was made in China for Telemania Products and distributed by Kash N' Gold Ltd., Ronkonkoma, New York. Model: Dino Telephone. 1996. 12.5" H x 5.25" W x 16" length. Touch-tone or pulse-dialing switch underneath base. Last-number redial key. Dino sound when phone rings. Ringer or dinosaur sound on/off switch underneath base. Requires two AA batteries. Keypad and key underneath rock lid on base. Transmitter and receiver located on handset back of dinosaur. On/off hook switch under handset. Desk set. (*Courtesy of Betty C.*) $20-30.

The Dodge Dakota Pickup Telephone was made in Taiwan for Unsonic. Model: CDT-129 NAB. Circa 1990s. 1.75" H x 9" length. Touch-tone or pulse-dialing switch. Last-number redial key. Mute key. Ringer on/off switch. Keypad, keys, switches, on/off hook switch, transmitter, and receiver underneath Dodge. Desk set or wall mount. (*Courtesy of Timothy J. Eisenhauer and Jonathan Frank.*) $10-20.

The Dodge Viper Telephone was manufactured in China for Telemania Products and Distributed by Kash N' Gold Ltd., Ronkonkoma, New York. 1994. 1.75" H x 7" length. Touch-tone or pulse-dialing switch. Last-number redial key. When phone rings headlights blink and horn honks. Keypad, key, switch, transmitter, receiver, and on/off hook switch underneath car. Desk set. (*Courtesy of Timothy J. Eisenhauer and Jonathan Frank.*) $15-25.

The Duck Telephone was made for Specialty Phones Incorporated. Model: MDT. 1983. 8" H x 5" W x 13.5" length. Touch-tone dialing. Quacks when phone rings. Keypad, on/off hook switch, transmitter, and receiver located on handset on top of ducks back. Desk set. (*Courtesy of Jim and Wynona Goodin.*) $50-60.

The Erotica Telephone was made in Hong Kong. Circa 1980s. 10.5" H x 4.5" W x 5" deep. Touch-tone dialing. Last-number redial key. Keypad, key, and on/off hook switch underneath base. Transmitter and receiver on front of figure. Desk set. $130-160.

The Fido Fone Dog with Bone Telephone was made in Hong Kong for Besco Electronic Manufacturing Co. Ltd. Model: NV 311. Circa 1990s. 11.5" H x 9" diameter. Touch-tone or pulse-dialing switch. Last-number redial key. Memory for auto dialing. Dog barks and eyes and mouth move when phone rings. Hi/Lo volume switch. Speaker phone. Keypad, key, switch, and on/off hook switch on top of base. Desk set. (*Courtesy of Jim and Wynona Goodin.*) $125-135.

18 PCT
Emergency Dial 911
POLICE

Opposite page:
Top: The '56 Thunderbird Car Telephone was made in China for Telemania Products and distributed by Kash N' Gold Ltd., Ronkonkoma, New York. Model: HAC '56 Thunderbird. 1996. 2.5" H x 2.75" W x 7.5" length. Touch-tone or pulse-dialing switch. Ringer on/off switch. Last-number redial key. Keypad, switches, key, on/off hook switch, transmitter, and receiver located underneath car. Desk set. (*Courtesy of Betty C.*) $35-45.
Center: The '57 Chevrolet Car Telephone was made in China for Telemania Products and distributed by Kash N' Gold Ltd., Ronkonkoma, New York. Model: HAC '57 Chevy. 1992. 2.5" H x 2.75" W x 7.5" length. Touch-tone or pulse-dialing switch. Ringer on/off switch. Last-number redial key. Keypad, switches, key, on/off hook switch, transmitter, and receiver located underneath car. Desk set. (*Courtesy of Betty C.*) $35-45.
Bottom: The '57 Chevrolet Police Car Telephone was made in China for Telemania Products and distributed by Kash N' Gold Ltd., Ronkonkoma, New York. Model: HAC '57 Chevy. Circa 1990s. 2.5" H x 2.75" W x 7.5" length. Touch-tone or pulse-dialing switch. Ringer on/off switch. Last-number redial key. Keypad, switches, key, on/off hook switch, transmitter, and receiver located underneath car. Desk set. (*Courtesy of Betty C.*) $35-45.

Right: The Fire 911 Box Telephone was made in China for Randix Corp. Incorporated. 1992. 16" H x 9.5" W x 7.25" deep. Touch-tone or pulse-dialing switch. Ringer on/off switch. Police, Fire, and Amcare emergency keys. Thirteen-number memory. Keypad, switch, keys, on/off hook switch, transmitter, and receiver located on handset. Desk set. (*Courtesy of Betty C.*) $100-110.

Below: The '57 Chevrolet Truck Telephone was made in China for Telemania Products and distributed by Kash N' Gold Ltd., Ronkonkoma, New York. Model: Pickup Truck. 1991. 3.75" H x 3.25" W x 8.5" length. Touch-tone or pulse-dialing switch. Last-number redial key. Keypad, switch, key, on/off hook switch, transmitter, and receiver located underneath. Desk set. (*Courtesy of Betty C.*) $35-45.

Below: The Fire Engine 1922 Telephone was made in China for Telemania Products and distributed by Kash N' Gold Ltd., Ronkonkoma, New York. Model: Fire Engine Telephone. 1996. 5" H x 4.5" W x 15" length. Touch-tone or pulse-dialing switch. Last-number redial key. Mute key. When phone rings engine starts and siren and bells sound. Ringer or Fire Engine sound switch. Phone on/off switch. Flash key for new dial tone. Requires two AA batteries. Keypad, switches, keys, on/off hook switch, transmitter, and receiver located on handset. Desk set. (*Courtesy of Betty C.*) $65-75.

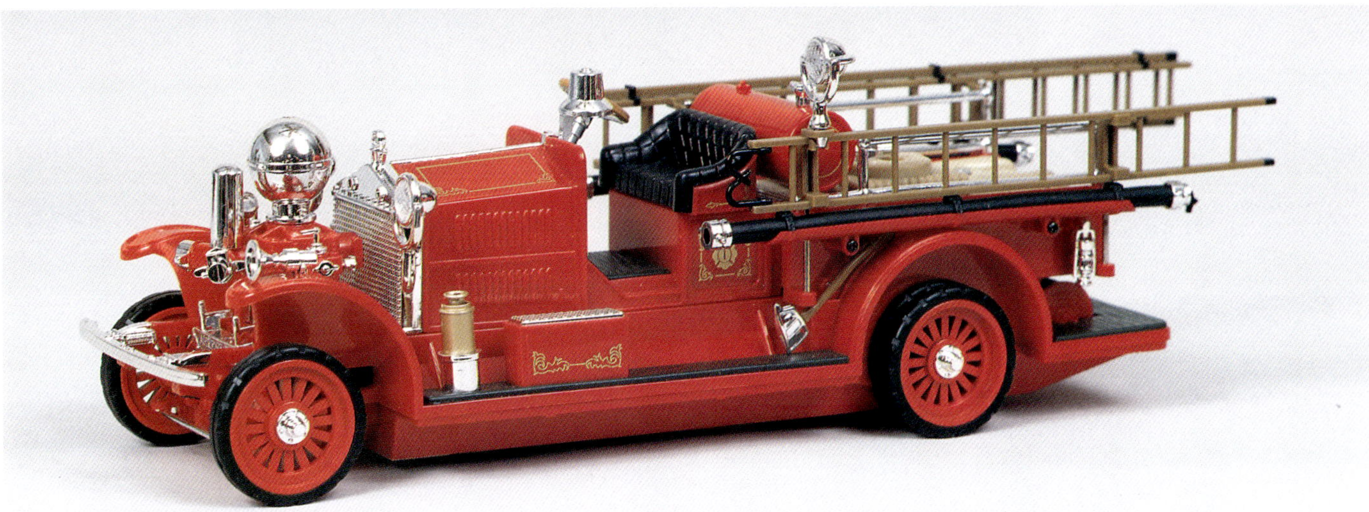

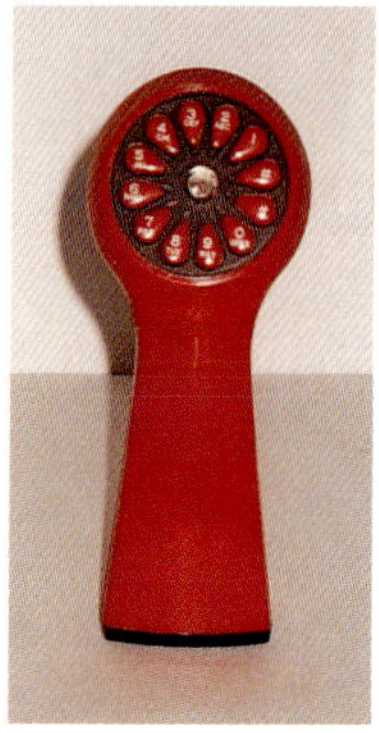

The Flower Telephone was made in Taiwan by Leetel. Model: KT-880P. Circa 1990s. 8" H x 3" W. Touch-tone dialing. Blinks when phone rings. Keypad, on/off hook switch, transmitter, and receiver on handset. Desk set. (*Courtesy of Timothy J. Eisenhauer and Jonathan Frank.*) $15-25.

The Flag Candlestick Telephone was made in the United States by American Telecommunications Corp., Elmonte, California. Circa 1970s. 12.5" H x 7" W x 5.25" deep. Pulse dialing. On/off hook switch handset holder. Transmitter mouthpiece. Receiver handset. Customer bought shell of phone and the telephone company installed dial and network. Desk set. $40-60.

The Frog Telephone was made in China. Circa 1990s. 3.5" H x 2.25" W x 5.5" length. Touch-tone dialing. Keypad, on/off hook switch, transmitter, and receiver underneath lid. Desk set. (*Courtesy of Betty C.*) $30-40.

The Golf Bag Telephone was made in China for Nicrotel Phone Corp. Circa 1990s. 7.5" H x 4.5" W. Touch-tone dialing. Last-number redial key. Mute key. Ringer on/off switch. LED in-use indicator. Keypad, keys, switch, on/off hook switch, transmitter, and receiver on handset. Desk set. (*Courtesy of Timothy J. Eisenhauer and Jonathan Frank.*) $30-50.

The Gumball Machine Telephone was made in Taiwan for Carousel Industries. Circa 1990s. 11.25" W x 7.5" diameter. Touch-tone dialing. Keypad and on/off hook switch transmitter, and receiver located on handset. Desk set. (*Courtesy of Betty C.*) $60-70.

The Grand Piano Telephone was made in China for Columbia TEL-COM Group Incorporated. Model: PN-10. Circa 1980s. 3.25" H x 7.75" W x 6.25" deep. Touch-tone or pulse-dialing switch. Last-number redial key. Mute key. Keypad, keys, switch, on/off hook switch, transmitter, and receiver on handset. Desk set. (*Courtesy of Betty C.*) $60-70.

The Happy Face Disc Telephone was manufactured in China for Telemania Products and distributed by Kash N' Gold Ltd., Ronkonkoma, New York. Model: Smiley Face. 1996. 3.25" deep 11.75" diameter. Touch-tone or pulse-dialing switch on side of handset. Ringer music/bell/off switch on back. Bell tone hi/low switch on back. Lights flash/on/off switch on back. Hearing aid compatible. Bell tone or music sounds and front lights up when phone rings. Star and pound keys for various purposes on long distance or other tone/dial system applications. Last-number redial key. Mute key. Flash key for new on/off hook. Keypad, keys, transmitter and receiver on handset. On/off hook switch in handset cradle. Desk set or wall mount. (*Courtesy of Betty C.*) $75-85.

The Heart Crusader Telephone was made in Taiwan for Tranbon Electronic Industrial Co. Inc. Model: Heart-2. Circa 1980s. 3.25" H x 6.75" W x 9" length. Touch-tone or pulse-dialing switch. Last-number redial key. Mute key. Keypad, switch, keys, on/off hook switch, transmitter, and receiver on handset. Desk set or wall mount. $20-40.

The Helicopter Telephone was made in China for Ronsonic Trading Corp., Old Bridge, New Jersey. Model: RON-125. Circa 1990s. 6" H x 4.25" W x 13.5" length. Touch-tone or pulse-dialing switch. Hearing aid compatible. Last-number redial key. Reset key for new dial tone. Volume hi/low control underneath base. On/off hook switch on top of chopper. Keypad, keys, switch, transmitter, and receiver on handset. Desk set. (*Courtesy of Betty C.*) $65-75.

The Hot Dog Telephone was made in China. Model: HD-8003. 1992. 2.25" H x 3" W x 8.25" length. Touch-tone or pulse-dialing switch. Last-number redial key. Mute key. Keypad, keys, switch, on/off hook switch, transmitter, and receiver underneath hot dog. Desk set. (*Courtesy of Betty C.*) $25-35.

The Hot Lips Telephone was made in China for Telemania Products and distributed by Kash N' Gold Ltd., Ronkonkoma, New York. Model: LIPS. Circa 1990s. 4" H x 8.5" W x 2.5" deep. Touch-tone or pulse-dialing switch. Last-number redial key. Keypad, key, switch, on/off hook switch, transmitter, and receiver located on handset top part of lips. Desk set. (*Courtesy of Betty C.*) $30-40.

The Hunting Dog Chest Telephone by Decotel. Circa 1970s. 5" H x 10.25" H x 8.25" deep. Touch-tone dialing. On/off hook switch in cradle underneath handset. Transmitter and receiver on handset. Keypad on top of base. Desk set. (*Courtesy of Betty C.*) $65-75.

The Jewelry Telephone was made in China for World Wide Design. Model: Love Phones SP-318S. 1993. 8.5" H x 2.5" H x 2.5" deep. Touch-tone dialing. Mute key. On/off switch underneath. Hearing aid compatible. Keypad, keys, on/off hook switch, transmitter, and receiver on front of handset. Desk set. (*Courtesy of Betty C.*) $25-35.

The Key Telephone was made in Hong Kong by Talk to Me. Model: KP-301. 1986. 1.25" H x 2.5" W x 9" length. Touch-tone dialing. Last-number redial key. Mute key. Ringer on/off switch. Keypad, keys, switch, on/off hook switch, transmitter, and receiver on handset. Desk set. (*Courtesy of Timothy J. Eisenhauer and Jonathan Frank.*) $10-25.

The Kitty Telephone was made in China for Minya International. Optional basket for Kitty to lie in. Model: KF-8007. Circa 1990s. 4.5" H x 4" W x 8" length. Touch-tone or pulse-dialing switch. Last-number redial key. Mute key. Keypad, switch, keys, on/off hook switch, transmitter, and receiver located underneath kitty. Desk set. (*Courtesy of Betty C.*) $25-35.

The Hello Kitty Telephone was made in Japan for Sanrio Co. Ltd. 1976. 10" H x 9" W x 8.25" deep. Pulse dialing. Dial and on/off hook switch on top of base. Ringer volume control underneath base. Transmitter and receiver on handset. Desk set. $30-40.

The Lamborgini Car Telephone was manufactured in China for Telemania Products and distributed by Kash N' Gold Ltd., Ronkonkoma, New York. 1994. 5.5" H x 7.25" W x 15" length. Touch-tone or pulse-dialing switch. Mute key. Last-number redial key. Requires four C-size batteries. Battery compartment under base. Hold key. Lights up and motor sounds when phone rings. On/off phone switch. Ringer or car "varoom" sound switch. Transmitter and receiver on handset. Switches and speaker underneath base. Keypad, keys, switches, and on/off hook switch, on handset under car. Desk set. (*Courtesy of Betty C.*) $80-90.

The Locomotive 1925 Crescent Train Telephone was made in China for Telemania Products and distributed by Kash N' Gold Ltd., Ronkonkoma, New York. Circa 1990s. 4.5" H x 3.5" W x 12.5" length. Touch-tone or pulse-dialing switch. Ringer or train sound on/off switch. Makes all sounds of a train. Last-number redial key. Mute key. On/off switch underneath base. Requires two AA batteries. Keypad, switches, key, on/off hook switch, transmitter, and receiver located underneath train. Desk set. (*Courtesy of Betty C.*) $85-95.

The London Booth Telephone was made in China for Polyconcept USA incorporated. Model: 541.055. 1996. 18.5" H x 8.5" W x 7.75" deep. Touch-tone or pulse-dialing switch inside booth. Transmitter and receiver located on handset inside booth. Light on/off switch inside booth. Requires 110V A/C for light inside booth. Keypad inside booth. On/off hook switch underneath handset inside booth. Desk set or wall mount. (*Courtesy of Betty C.*) $75-85.

The Met-a-Phone is an early example of a Novelty Telephone. 1901. This art deco type phone was made with an antique holder for the handset. Some were made with a pen holder. (*Courtesy of Ken King.*) Priceless.

The Ocean Racer Speedboat Telephone was made in China for Ronsonic Trading Corp., Old Bridge, New Jersey. Model: HAC-RON-135. Circa 1990s. 4" H x 3.5" W x 13.25" length. Touch-tone or pulse-dialing switch. Last-number redial key. Hearing aid compatible. Mute key. Flash key for new dial tone. Racer sound or ringer on/off switch. Keypad, keys, switches, on/off hook switch, transmitter, and receiver on handset top of racer. Desk set. (*Courtesy of Betty C.*) $70-80.

The Orange Telephone was made in China. 1992. 4.25" H x 4.5" diameter. Touch-tone or pulse-dialing switch. Last-number redial key. Mute key. Keypad, switch, keys, on/off hook switch, transmitter, and receiver underneath lid. Desk set. (*Courtesy of Betty C.*) $40-50.

The Organ Telephone was made in Taiwan. Model: HAC-SH-121 (DT-606). 1992. 4" H x 9.75" W x 4.25" deep. Touch-tone or pulse-dialing switch and ringer hi/low switch on back of organ. Keypad organ keys. On/off hook switch in cradle under handset. Transmitter and receiver on handset. Desk set or wall mount. (*Courtesy of Betty C.*) $50-60.

Below: The P51D Mustang Fighter Airplane Telephone was made in China for Telemania Products and distributed by Kash N' Gold Ltd., Ronkonkoma, New York. Model: Fighter Plane Telephone. 1995. 4.5" H x 10.5" W x 12.5" length. Touch-tone or pulse-dialing switch. Last-number redial key. When phone rings plane makes "vroom-vroom" sound. Ringer or plane sound switch. Requires two AA batteries. Keypad, switches, key, on/off hook switch, transmitter, and receiver located underneath plane. Desk set. (*Courtesy of Betty C.*) $65-75.

Above: The Panda Bear Telephone was made in Taiwan. Model: HAC-BA-222. 1996. 3.5" H x 4.5" diameter. Touch-tone or pulse-dialing switch. Last-number redial key. Keypad, switch, key, on/off hook switch, transmitter, and receiver underneath bear. Desk set. (*Courtesy of Betty C.*) $35-45.
Below: The Pig Telephone was made in Taiwan for Youmax S & T Inc., Walnut, California. Model: PG-666. Circa 1990s. 3.25" H x 4.75" H x 8" length. Touch-tone or pulse-dialing switch. Last-number redial. Keypad, key, switch, on/off hook switch, transmitter, and receiver underneath pig. Desk set. (*Courtesy of Betty C.*) $25-35.

The Pig Phone Telephone was made in Taiwan for Telemania Products and distributed by Kash N' Gold Ltd., Ronkonkoma, New York. Model: HAC PG-666. Circa 1990s. 10.75" H x 5.75" W x 4.5" deep. Touch-tone dialing. Pig snores when phone rings. Requires 9V battery compartment underneath. Keypad, on/off hook switch, transmitter, and receiver located on handset. Desk set. (*Courtesy of Betty C.*) $25-35.

The Porsche Car Telephone. Circa 1990s. 2" H x 3" H x 5.25" length. Touch-tone dialing. Keypad, on/off hook switch, transmitter, and receiver underneath car. Desk set. (*Courtesy of Betty C.*) $15-25.

The Puppy Telephone was made in China Minya International. Optional basket for puppy to lie in. Model: PF-8006. 1994. 2.5" H x 4.75" W x 8.5" length. Touch-tone or pulse-dialing switch. Last-number redial kcy. Mute key. Keypad, switch, keys, on/off hook switch, transmitter, and receiver located underneath puppy. Desk set. (*Courtesy of Betty C.*) $30-40.

The Purse Telephone was made in China. Sold by Shoe Warehouse, Midwest City, Oklahoma. Circa 1990s. 11" H x 10" W x 3" deep. Touch-tone dialing. Ringer on/off switch. Keypad, switch, on front of purse. On/off hook switch, transmitter, and receiver on hand set inside purse. Desk set or wall hanger. (*Courtesy of Betty C.*) $85-95.

The Quacker Duck Telephone was made in Taiwan for Telemania Products and distributed by Kash N' Gold Ltd., Ronkonkoma, New York. 1991. 4.5" H x 3.25" W x 10" length. Touch-tone or pulse-dialing switch. Last-number redial key. Duck quacks when phone rings. Keypad, key, switch, on/off hook switch, transmitter, and receiver located underneath duck. Desk set. (*Courtesy of Betty C.*) $65-75.

The Santa Claus Telephone was made in China. 1996. 14" H x 8.25" W x 7" deep. Touch-tone or pulse-dialing switch. Ringer or music switch. When phone rings plays "Here Comes Santa Claus" and lights up. Requires batteries. Keypad, switches, on/off hook switch, transmitter, and receiver located on handset. Desk set. (*Courtesy of Betty C.*) $60-70.

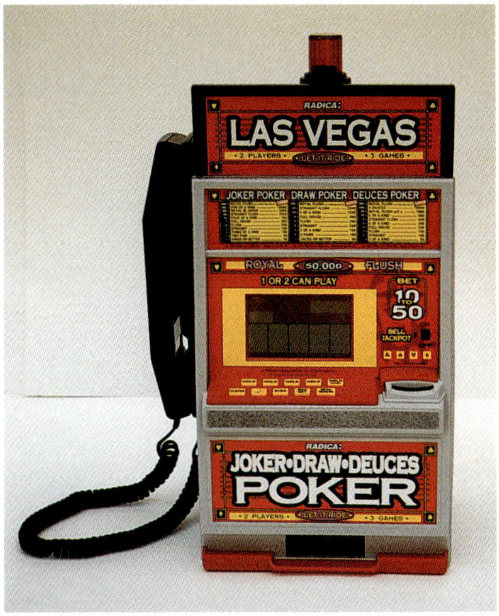

The Slot Machine Telephone was made in China for Radica. Model: 3119G3. Circa 1980s. 14" H x 8.5" W. Touch-tone or pulse-dialing switch. Bell on/off switch. Keypad, switches, on/off hook switch, transmitter, and receiver on handset. Desk set. (*Courtesy of Timothy J. Eisenhauer and Jonathan Frank.*) $40-60.

The Soccer Ball Sports Fone Telephone was made in China for Minya International. Model: SC 392. Circa 1990s. 5" H x 4.25" diameter. Touch-tone or pulse-dialing switch. Last-number redial key. Mute key. Keypad, switch, keys, on/off hook switch, transmitter, and receiver located underneath lid. Desk set. (*Courtesy of Betty C.*) $40-50.

The Teddy Bear Telephone was made by Teleconcepts Incorporated. Model: 341542. 1991. 10" H x 7" W x 7" deep. Touch-tone or pulse-dialing switch. Last-number redial key. Ringer on/off switch. LED light and on/off switch. Keypad, key, switches, on/off hook switch, transmitter, and receiver on handset on back of bear. Desk set. (*Courtesy of Betty C.*) $30-40.

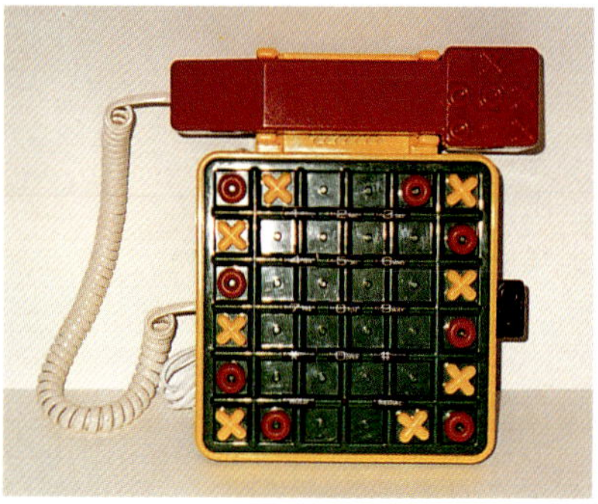

The Tic Tac Toe Telephone was made in Taiwan for Le Tel. Model: DT-777. 1994. 2.5" H x 6" W x 9" length. Touch-tone dialing. Last-number redial key. Ringer hi/low switch. Sliding X and O drawer. Keypad, keys, on/off hook switch, transmitter, and receiver underneath lid. Desk set. (*Courtesy of Timothy J. Eisenhauer and Jonathan Frank.*) $25-50.

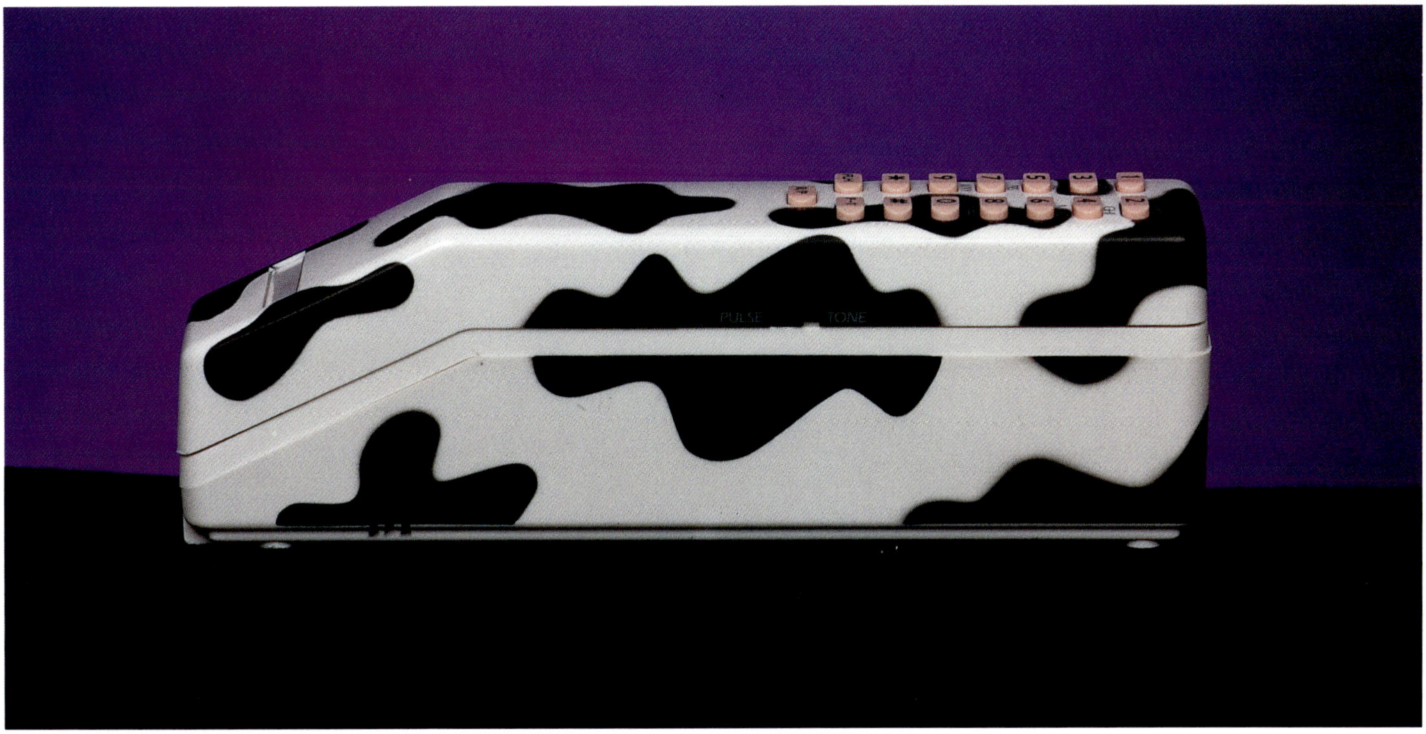

The Tiffany Cow Telephone was manufactured in China and distributed by Kash N' Gold Ltd., New York, under license from the Coca-Cola Co. Model: HC 900/Tiffany. 1994. 2.5" H x 1.5" W x 7.25" length. Touch-tone or pulse-dialing switch. Last-number redial R/P key. Flash FLH key for new dial tone or call waiting or three party calling. P-T key for rotary subscribers to change from pulse to tone mode for low cost long distance and computer services. Keypad, keys, switch, on/off hook switch, transmitter, and receiver on handset. Desk set or wall mount. (*Courtesy of Betty C.*) $35-45.

The Tomato Telephone was made in China. 1992. 3.5" H x 4.5" diameter. Touch-tone or pulse-dialing switch. Keypad, switch, on/off hook switch, transmitter, and receiver underneath lid. Desk set. (*Courtesy of Betty C.*) $30-40.

The Upright Piano Telephone was made in China for KCL Technology Ltd. 1990. 8" H x 7.75" W x 5" deep. Touch-tone or pulse-dialing switch. Last-number redial key. Plays "When The Saints Go Marching In" when phone rings. Melody or ring on/off switch. Requires two AA batteries. Keypad, key, switches, on/off hook switch, transmitter, and receiver on hand set. Desk set. (*Courtesy of Betty C.*) $85-100.

Opposite page:
Bottom: The Volkswagon Convertible Car Telephone was made in China for Polyconcept USA Inc. Model: PFA Beetle Phone. 1996. 4" H x 5" W x 10" length. Touch-tone or pulse-dialing switch. Ringer on/off switch. Last-number redial key. Flash key for new dial tone. Lights flash when phone rings. Requires two AA batteries. Keypad, switches, key, on/off hook switch, transmitter, and receiver located underneath car. Desk set. (*Courtesy of Betty C.*) $50-65.

Above: The Toucan Bird Telephone was made in China. Model: T-01. Circa 1990s. 6.25" H x 2.75" W x 6" deep. Touch-tone or pulse-dialing switch. Hearing aid compatible. Keypad, switch, on/off hook switch, transmitter, and receiver located inside. Desk set. (*Courtesy of Betty C.*) $30-40.

Right: The Violin Telephone was made in China for Ronsonic Trading Corp., Old Bridge, New Jersey. Model: RON-108. Circa 1990s. 11.75" H x 4.75" W x 8" deep. Touch-tone or pulse-dialing switch. Hearing aid compatible. Last-number redial key. Pause or mute key. Flash key for new dial tone. On/off hook switch underneath handset. Keypad, keys, switch, transmitter, and receiver on handset. Desk set. (*Courtesy of Betty C.*) $60-70.

The World Globe Telephone was made in China by The World Fone. Model: TW-393. 1993. 4" diameter. Touch-tone dialing. Last-number redial key. Mute key. Keypad, keys, on/off hook switch, transmitter, and receiver inside globe. Desk set. *(Courtesy of Timothy J. Eisenhauer and Jonathan Frank.)* $20-35.

The World War II Field Speaker Telephone was made in China for Polyconcepts USA Incorporated. Made for SOSS Collection Spirit of St. Louis. 1995. 9.5" H x 10" W x 4" deep. Touch-tone or pulse-dialing switch underneath. Last-number redial key. Ten number memory recall key. Store numbers key. Flash key for new dial tone. Speaker and ringer volume controls. Speaker phone key. Ringer light on/off key. Speaker in use light. Keypad, keys, and speaker on top of base. Transmitter and receiver on handset. Desk set or wall mount. *(Courtesy of Betty C.)* $40-60.

Chapter IV

ITSY BITSY TEENY WEENY

Miniature Telephones

By Ken King

One of the most desirable miniature collectibles among telephone collectors and general antique collectors are reduced-scale copies of actual telephones. The little telephones usually measure between 2.5 and 3.5 inches in height, although a few will vary by a few inches.

Miniature telephones are commonly referred to as "salesman samples" due to their exacting detail and conformity to the full-size working telephones used throughout America. However, research and conversations with early telephone pioneers generally conclude that the original factory models produced from 1910 to 1915 were gifts given to large and small telephone exchange owners, the commercial buyers of working telephones and supplies.

The most common factory produced miniature is the early 1900s model of the Stromberg-Carlson desk stand. This model is typically marked with the company name across the top of the transmitter cup. This same model was also less frequently marked on the face of the transmitter with the name "Home Telephone Company." A much less common, factory produced version was modeled after the Dean or Garford Telephone Company desk stand of the same era. It was almost always marked on the face of the transmitter. This version appears to have had more varied name markings than the Stromberg-Carlson model.

Beginning in 1988 a long-time antique telephone collector, who is both a skilled artisan and craftsman, began very limited miniature production of many of the popular telephones used in America. These models are each hand-made and were generally limited to only four of each phone. The first models produced were not marked. After the initial unmarked models, the later productions were stamped "E. B. Ellis, Maker." (The word "Maker" appears under E. B. Ellis.) More than forty-five different miniature telephones have been produced. Among the rarest of the Ellis miniatures are three models that actually work. The most recent technology allows very detailed markings on the transmitter face.

The value of the factory produced miniatures and the more recent Ellis productions continues to increase. The typical value for a complete miniature in good condition will bring from $100 to $250. Some of the rare models may command even higher prices.

Four factory miniatures left to right: Stromberg-Carlson, Garford Telephone Co., Kansas City Home, and Home Telephone Co. See following page. (*Courtesy of Ken King.*)

Stromberg-Carlson factory miniature with markings on transmitter cup. (*Courtesy of Ken King.*)

Stromberg-Carlson factory miniature marked "Home Telephone Co." (*Courtesy of Ken King.*)

Garford-Dean Manufacturing Co. factory miniature with transmitter marked. (*Courtesy of Ken King.*)

Garford-Dean Electric Co. factory miniature marked "Kansas City Home." (*Courtesy of Ken King.*)

Transmitter markings of Garford-Dean factory miniature. (*Courtesy of Ken King.*)

American Electric Co. candlestick. (E. B. Ellis) (*Courtesy of Ken King.*)

Swedish-American factory miniature 1.25 inches. (Note: base view in picture below. (*Courtesy of Ken King.*)

Swedish-American Telephone Co., Chicago. (view of base) (*Courtesy of Ken King.*)

American Electric Co. fluted shaft. (E. B. Ellis) (*Courtesy of Ken King.*)

American Electric Co. pot belly. (E. B. Ellis) (*Courtesy of Ken King.*)

Bell Laboratories factory miniature. (*Courtesy of Ken King.*)

Bell Telephone Co., Canadian No. 4. (E. B. Ellis) (*Courtesy of Ken King.*)

B&R Electric and Telephone Mfg. Co. tapered shaft. (E. B. Ellis) (*Courtesy of Ken King.*)

Century Telephone Construction Co. candlestick. (E. B. Ellis) (*Courtesy of Ken King.*)

Chicago Telephone Supply Co. Sears-Roebuck candlestick. (E. B. Ellis) (*Courtesy of Ken King.*)

Chisholm-Dunn Telephone Appliances Ltd. candlestick. (E. B. Ellis) (*Courtesy of Ken King.*)

Clark Automatic Telephone Automatic Switchboard Co. miniature. Shown with actual size telephone. (E. B. Ellis) (*Courtesy of Ken King.*)

Clark Automatic miniature. (E. B. Ellis) (*Courtesy of Ken King.*)

Connecticut Telephone and Electric Co. candlestick. (E. B. Ellis) (*Courtesy of Ken King.*)

Couch and Seeley pot belly. (E. B. Ellis) (*Courtesy of Ken King.*)

Dean-Garford pot belly. (E. B. Ellis) (*Courtesy of Ken King.*)

Garford-Dean candlestick. (E. B. Ellis) (*Courtesy of Ken King.*)

Eastern Telephone Manufacturing Co. tapered shaft. (E. B. Ellis) (*Courtesy of Ken King.*)

Electric Goods Manufacturing Co. rope shaft. (E. B. Ellis) (*Courtesy of Ken King.*)

Holtzer-Cabot Telephone Co. pot belly wood base. (E. B. Ellis) (*Courtesy of Ken King.*)

Julias Andrae candlestick. (E. B. Ellis) (*Courtesy of Ken King.*)

Kellogg Switchboard and Supply Co. candlestick. (E. B. Ellis) (*Courtesy of Ken King.*)

Mianus Electric Co. tapered shaft. (E. B. Ellis) (*Courtesy of Ken King.*)

North Electric Co. pot belly. (E. B. Ellis) (*Courtesy of Ken King.*)

Standard Telephone and Electric Co. rope shaft wood base. (E. B. Ellis) (*Courtesy of Ken King.*)

Sterling Electric Co. fluted shaft. (E. B. Ellis) (*Courtesy of Ken King.*)

Stromberg-Carlson 1897 Roman Column. (E. B. Ellis) (*Courtesy of Ken King.*)

Stromberg-Carlson Telephone Manufacturing Co. 1897 Roman Column four inch working miniature. (E. B. Ellis) (*Courtesy of Ken King.*)

Stromberg-Carlson flip top 1896. (E. B. Ellis) (*Courtesy of Ken King.*)

Stromberg-Carlson tapered oil can candlestick 1899. (E. B. Ellis) (*Courtesy of Ken King.*)

Factory paperweights of telephone receivers. Left: Post and Co. and right: Stromberg-Carlson. (*Courtesy of Ken King.*)

Generic Dial cradle phone paperweight. (*Courtesy of Ken King.*)

Factory Paperweights from left: Bell System 300 set and generic dial cradle. (*Courtesy of Ken King.*)

Pocket watch chain fobs representing from left to right: Swedish-American, American Electric, and Stromberg-Carlson. (*Courtesy of Ken King.*)

Strowger Automatic dial pot belly. (E. B. Ellis) (*Courtesy of Ken King.*)

Swedish-American candlestick. (E. B. Ellis) (*Courtesy of Ken King.*)

Utica Fire Alarm Telephone Co. pot belly three inch working miniature. (E. B. Ellis) (*Courtesy of Ken King.*)

Viaduct Manufacturing Co. square base with rope shaft. (E. B. Ellis) (*Courtesy of Ken King.*)

Viaduct Manufacturing Co. candlestick. (E. B. Ellis) (*Courtesy of Ken King.*)

Viaduct Manufacturing Co. tapered shaft. (E. B. Ellis) (*Courtesy of Ken King.*)

Wood wall telephone miniatures. Right: Western Electric fiddleback and left: Chicago Telephone Co. picture frame cathedral-top wall phone. (E. B. Ellis) (*Courtesy of Ken King*.)

Western Electric pot belly 1897 with marked transmitter. (E. B. Ellis) Note: See picture below. (*Courtesy of Ken King*.)

Western Electric Co. 1897 pot belly 4-inch working miniature. (E. B. Ellis) (*Courtesy of Ken King*.)

Western Electric l897 pot belly view of transmitter. (E. B. Ellis) (*Courtesy of Ken King*.)

Western Electric dial candlestick. (E. B. Ellis) (*Courtesy of Ken King.*)

Western Electric No. 10. (E. B. Ellis) (*Courtesy of Ken King.*)

Western Electric 1900 type. (E. B. Ellis) (*Courtesy of Ken King.*)

Western Electric B&R pot belly. (E. B. Ellis) (*Courtesy of Ken King*.)

Western Electric 1897 pot belly. (E. B. Ellis) (*Courtesy of Ken King*.)

Wilhelm Telephone Mfg. Co. fluted shaft. (E. B. Ellis) (*Courtesy of Ken King*.)

Chapter V

BIG BOYS' TOYS

Toy Telephones

Because of the increasing cost and scarcity of collectible telephones, many antique lovers seek other ways to satisfy their collecting habit. Toy telephones are in demand and have steadily increased in value.

The toy phones are in the form of candlesticks, desk sets, pay phones, and switchboards. These phones are made of a variety of materials: metal, wood, glass, and plastic. They usually have strings for cords.

Children have continued to be charmed by the ring of a telephone. Collectors can count on a wide variety of new toy phones to whet their collecting appetite.

Chrome Switchboard requires one D battery 10.5"H. $75-100.

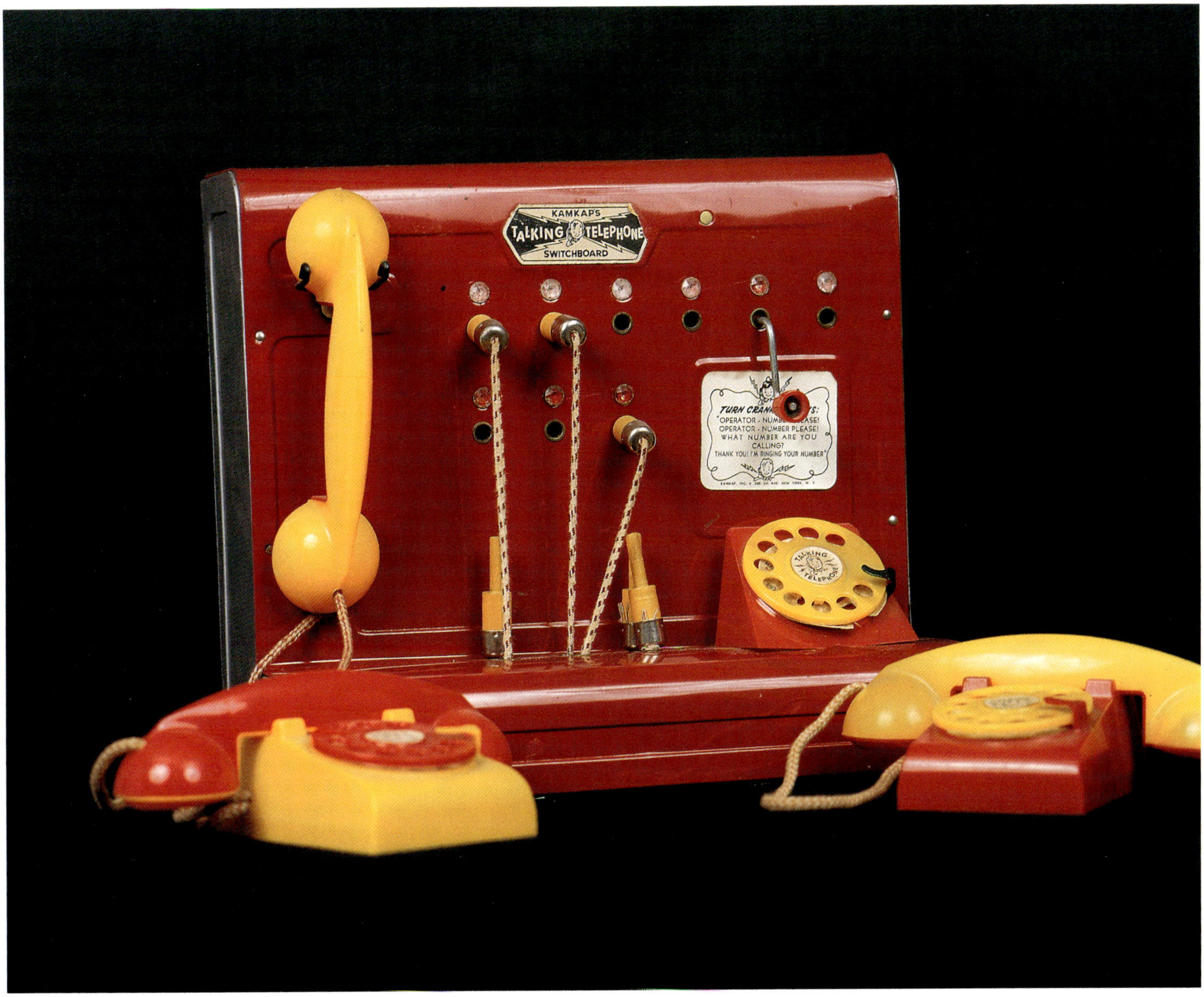

Switchboard Talking Telephone ,10.5" H, by Kamkap Inc., New York. $75-100.

TOLL

L to R: dial Candlesticks, PLA Phone-7200, 7" H. Playphone-0635, 7.25" H. Manufactured by The Gong Bell Manufacturing, East Hampton, Connecticut. $50-70.

Opposite page:
The Candlestick Bell Phone Toll Bank was manufactured by the N.N. Hill Brass Co., East Hampton, Connecticut. Model: Kiddie 9432. 8" H. $65-85.

L to R: Candlestick with Bells Playphone-600, 7.75" H, June 22, 1922. Junior Playphone-500, 7" H, June 22, 1922. Silver Chime Playphone-625, 8.55" H. Junior Playphone, 6.5" H. All manufactured by The Gong Bell Manufacturing, East Hampton, Connecticut. $50-70.

Green Oval-base Candlestick with dial, 8" H. $40-50.

L to R: Candy Two-glass Candlestick containers 4.75"H. Victory Glass Co. Jeanette, Pennsylvania. Glass Deskset Candy Container, 3.75" H. E. Rozen Co., Pawtucket, Rhode Island. Plastic Candlestick Container with Candy. $20-45.

Marvel Talking phones set. Need four NO. 6 dry cells, metal, 3.5" H, Circa 1950s by Century Chicago Co., Chicago 22, Illinois. $90-125.

Plastic Candlestick, 9.5" H. Bell in base. $20-35.

L to R: Voicephone, 3.75" H, metal. Speedphone 3.5" H, metal. Both manufactured by The Gong Bell Co., East Hampton, Connecticut. $50-70.

L to R: Desk Set. Actual photographs in middle of dial changes for each number on dial oval base, 5" H, metal. Red metal desk set round base, 3.75" H. Desk set with alpha prefix on dial. 3.75" H. $45-65.

A.T. & T. Wire-Intercom telephone set. Requires two 9-volt batteries, 2.5" H, plastic. $35-50.

Mattel-O-Phone. Plays four-inch records, requires two 1.5-volt batteries, 1965, 4.75" H, plastic, by Mattel Inc. $30-50.

L to R: Blue Princess, plastic, 3.5" H. Aqua Princess, plastic, 2" H. Pink Princess, plastic, 3.5" H. $10-20.

L to R: Coin telephones by Ideal. 3.75" H. PLA-Pay Phone by The Gong Bell Manufacture, East Hampton, Connecticut. $30-50.

Kanter Wall Phone, Circa 1890s. (*Courtesy of Ken King*.) $125-175.

Metal Wall Phone Bank. Public Pay Telephone. 7.25" H. $30-50.

Tootsie Toy Deskset with Box. (larger than actual size) (*Courtesy of Ken King.*) $75-125.

L to R: Snoopy Romper Room, 1966, 10" H, plastic by Hasbro. Smurf, 1982, 9" H, plastic by H. G. Toys Inc., Long Beach, New York. $15-30.

Plastic Clown wall phone, 8.25" H, by Mascon. $10-20.

Bugs Bunny Talking Playphone, made in Singapore for Texas Instruments, 1993. $15-30.

Teddy Ruxpin by Worlds of Wonder. China, 1990. $10-20.

BIBLIOGRAPHY

Antique Trader Weekly, Antique Trader Publications, Dubuque, Iowa.

Dooner, Kate E., *Telephones Antique to Modern*, Schiffer Publishing, Ltd. West Chester, Pennsylvania.

Dooner, Kate E., *Telephone Collecting: Seven Decades of Design*, Schiffer Publishing, Ltd., Atglen, Pennsylvania.

Kovel, Ralph and Terry, *Kovel's Antiques & Collectibles Price Lists*, 1994-1997, Crown Trade Paperbacks, New York, New York.

Toy Shop, Krause Publications, Iola, Wisconsin.

Toy Trader, Antique Trader Publications, Dubuque, Iowa.

Warman's Today's Collector, Krause Publications, Iola, Wisconsin.